The Natural History of
INSECTS

THE NATURAL HISTORY OF INSECTS

ROD AND KEN PRESTON-MAFHAM

The Crowood Press

First published in 1996 by
The Crowood Press Ltd
Ramsbury, Marlborough
Wiltshire SN8 2HR

British Library Cataloguing-in-Publication Data
A catalogue record for this book is available from the British Library.

ISBN 1 85223 964 6

Title page: Parasitic fly, *Adejeania vexatrix*

All photographs by the authors, Premaphotos Wildlife.
Line illustrations by Alan Rollason.

Typeset and designed by:
D & N Publishing
DTP and Editorial Services
Ramsbury, Marlborough
Wiltshire SN8 2 HR

Typefaces used: main text, New Baskerville; captions, News Gothic; headings, News Gothic.

Printed in Great Britain by BPC Books Limited, Aylesbury
A member of The British Printing Company Limited

CONTENTS

01 Introduction to Insects

CLASSIFICATION

Where precisely in the animal kingdom do those familiar creatures, the insects, fit? There is an assemblage of groups, collectively known as the invertebrate animals, which have in common their lack of an internal bone or cartilage skeleton. Within the invertebrates are found such animals as the worms, the molluscs (slugs, snails, etc.) and the arthropods. It is within this last group or superphylum of animals that the insects are included. The basic characteristics of the major arthropod groups are as shown in the panel.

MAJOR ARTHROPOD GROUPS

Crustacea (crabs, lobsters, prawns, shrimps and woodlice) have five pairs of walking appendages, the first pair may be armed with pincers to assist in manipulating food.

Arachnida (spiders, scorpions, solifuges, harvestmen, whip-scorpions and pseudoscorpions) have four pairs of walking legs. (Pincers in scorpions, pseudoscorpions and whip-scorpions are on modified head appendages.)

Myriapoda have many pairs of walking legs, the centipedes with one pair per body segment and the millipedes with two pairs per body segment.

Insecta have three pairs of walking legs and two pairs (one pair in flies) of wings.

The darter dragonfly, **Trithemis ellenbecki**, from Kenya clearly shows most of the main insect external characteristics, that is, body divided into head, thorax and abdomen, a pair of legs on each thoracic segment and a pair of wings on the mesothorax and metathorax. This particular individual has adopted the 'obelisk' position, with its abdomen pointing directly towards the sun, a ploy which helps keep it cool in the heat of the day.

CLASSIFICATION OF THE HIVE BEE

GROUP	GROUP NAME	INCLUDES
Kingdom	Animalia	All animals
Superphylum	Arthropoda	All animals with an exoskeleton
Phylum	Uniramia	Those arthropods with a single pair of antennae and unbranched legs and other appendages
Superclass	Hexapoda	Those Uniramia with three pairs of walking legs
Class	Insecta	All hexapods except for the Protura, Diplura and Collembola
Order	Hymenoptera	Sawflies, wood wasps, bees, ants and wasps
Suborder	Apocrita	Bees, ants and wasps
Family	Apidae	Highly social bees
Genus	*Apis*	True honey bees
Species	*mellifera* ssp *carnica*	European honey bee

THE CLASSIFICATION OF THE SUPERPHYLUM ARTHROPODA

Phylum Chelicerata

CLASS ARACHNIDA
spiders, scorpions, ticks, mites, etc.

CLASS MEROSTOMATA
king-crabs or horseshoe-crabs

CLASS PYCNOGONIDA
sea-spiders

Phylum Crustacea
crabs, lobsters, shrimps, prawns, wood lice, etc.

Phylum Onychophora
velvet worms

Phylum Uniramia

SUPERCLASS MYRIAPODA

CLASS CHILOPODA
millipedes

CLASS DIPLOPODA
centipedes

SUPERCLASS HEXAPODA

CLASS PROTURA

CLASS DIPLURA

CLASS COLLEMBOLA
springtails

CLASS INSECTA
bugs, beetles, flies, etc.

Arthropods have in common a tough external skeleton (exoskeleton; *see* page 21) within which the body musculature and organs are contained. The subdivisions within the arthropods have changed somewhat in recent years and the classification shown in the box is now generally accepted by most taxonomists and entomologists.

It can be seen that classification of living organisms involves a hierarchy, with the superphylum Arthropoda divided into several phyla, each phylum then subdivided into superclasses, further subdivided into classes, and so on. Just to make it clear how the system works and before going on to discuss the arthropods and insects in greater detail, the table on the previous page shows the full classification of a single organism, the common domestic European form of the hive bee, *Apis mellifera carnica.*

Further subgroupings such as subclass and tribe are available for use where a greater subdivision is required.

WHAT IS AN INSECT?

Ask almost anyone what an insect is and they will say that it is an animal with six legs and wings. This is partly true, in that all insects have six pairs of legs, though these may be absent in some larval stages and modified in various ways in some adult insects, such as certain nymphalid butterflies. Furthermore, the most primitive insect groups lack wings. The possession of three pairs of walking legs does distinguish the insects from the spiders and their allies, which have four pairs, the crustacea, which have five pairs and the myriapods, which have many pairs. The Collembola (the springtails), along with the Protura and Diplura, also have three pairs of walking legs but lack wings. They were once thought of as insects but on account of certain primitive character they are no longer considered to be so, though they are closely related to insects and are included within the Hexapoda.

The illustration in Chapter 2 shows the main external character of an insect. All insects have a normally rigid exoskeleton, which contrasts with the internal skeleton (endoskeleton) of the vertebrate animals. In the vertebrates, the muscle systems are outside the skeleton, whereas in the arthropods, the skeleton basically consists of a series of rigid tubes and series of plates with the muscles running within the tubes and between the plates. In order for there to be some flexibility, to allow movement of the limbs and bending of the body, the exoskeleton is much thinner where joints occur between the plates and tubes. This is only a very simplistic view of the insect exoskeleton, which in reality is much more complex. Internally, there are many inward protrusions of the exoskeleton to allow for the attachment of the muscles.

One measure of the success of the insects is the sheer numbers of them which are in existence. Here, for instance, a group of butterflies are drinking on a river bank in Malaysia. This shows just a few of the hundreds that were there, close examination revealing that members of at least four different species were present.

INSECT SUCCESS

Without doubt, the insects are one of the most successful and diverse groups of living organisms in the world today. Estimates vary, but the number of different species in existence certainly exceeds one million and may run into several millions. Thousands of new species are discovered and described each year but in every entomological institution around the world there are many thousands of others awaiting their first description. Sadly, a proportion of these will almost certainly be extinct before they are even described, simply because their habitats will have been destroyed by man since their initial discovery. Furthermore, this same habitat destruction is almost certainly, at this very moment, wiping out species of insect which we have never even discovered. Although we humans think of ourselves as multitudinous, our numbers come nowhere near those of the insects, for it has been estimated that for each human alive at this moment there are 200 million insects.

This ichneumon wasp-mimicking coreid bug belonging to the genus **Holhymenia** was new to science when it was photographed in a Peruvian rainforest. Even today, and probably for the foreseeable future, as long as we do not destroy them, the world's rainforests will reveal thousands more species of insect as yet unknown to us.

What accounts for this enormous success of the insects? There are three main factors:

- THE EXOSKELETON
- SIZE
- DIET

Undoubtedly one of the greatest contributions to the success of insects is their chitinous exoskeleton (*see* page 21). This provides them with a tough outer layer which protects them against physical damage as well as preventing excess water loss, an invaluable asset to any terrestrial animal, and the majority of insects are terrestrial.

A second important factor in the success of insects is, interestingly, their small size. This allows many of them to live in a relatively small area where they can occupy spaces which are much too small for most vertebrate animals. Although bigger insects have existed prehistorically, the biggest insects in existence today are beetles weighing around 70g but these are exceptional with the majority of species much smaller than this. At the other extreme, the smallest insects are microscopic parasitic wasps less than 0.2mm in length. Small size also allows for a more rapid growth to maturity and a greater reproductive rate.

A third important factor contributing to the success of insects is their diet. There is virtually nothing of living origin which is not eaten by one species of insect or another. The success of the higher insects is added to by the fact that the adults and larvae exploit different food sources in most instances. It is with insect larvae that we find some of the most catholic tastes. Take the larvae scuttle flies of the family Phoridae as an example. The larvae of *Megaselia* species have been found feeding on such diverse substances as curdled milk, old cheese, human excrement, concentrated soap solution, shoe polish, emulsion paint, human corpses embalmed in formalin and the lungs of living humans, although any kind of decaying plant or animal material will do just as well. Even more bizarre are larvae of the ephydrid fly *Psilopa petrolei*, the petroleum fly. They live in, and feed on, the pools of crude oil which seep to the surface in the oil fields of California. They do not actually feed on the oil but on the dead animals and plants which fall into it.

02 Structure and Physiology of Insects

To possess a full understanding of the life of an insect it is necessary to know something of its structure (morphology) and the way in which the various body systems function (physiology).

A view of a wasp to show the main external characteristics of an insect.

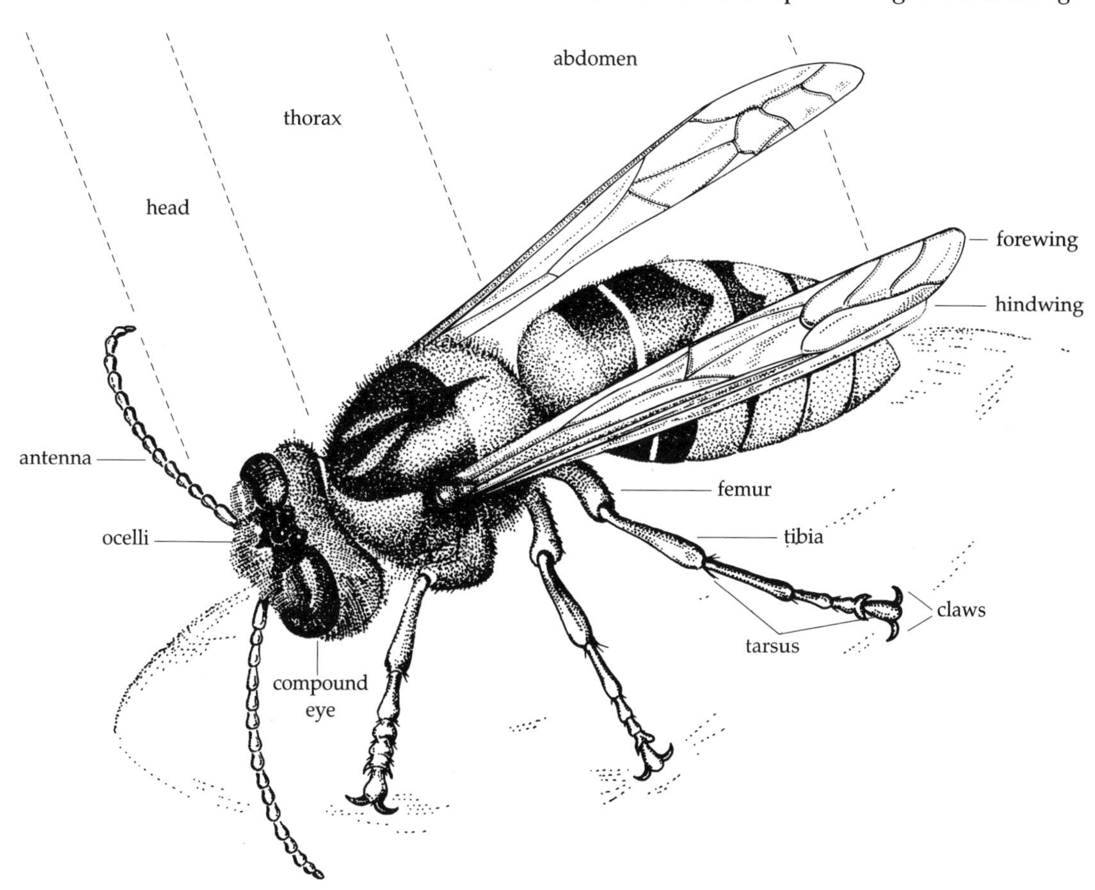

EXTERNAL STRUCTURE OF INSECTS

The body of adult insects is most obviously divided into three distinct sections, the head followed by the thorax and then the abdomen. The insect body is also divided up into segments although this appears most obvious when looking at the abdomen of an adult or an insect larva such as a caterpillar. The head and thorax are segmented, as we shall see, but less obviously so. Since there are a number of fundamental differences between the adult insect and the larva, the latter will be dealt with separately in a later section.

THE HEAD

The insect head is formed from a number of plates which are fused together to form a tough head capsule equivalent to the vertebrate skull. The relationship between the head and the thorax can take on one of two broadly different forms. The longitudinal axis of the head can be at right angles to the longitudinal axis of the thorax, when it is referred to as a hypognathous head. Alternatively, it can lie in the same plane as the thorax, or tipped slightly down, when is is called a prognathous head. These two different forms of head affect the plane in which the jaws work. In a number of plant-feeding insects, such as aphids and cicadas, belonging to the suborder Homoptera, the head is inclined so that the sucking mouthparts point backwards (an opisthognathous head).

The head bears a number of structures concerned with feeding and with sensory perception. Most obviously, the head of the majority of insects has a pair of compound eyes, though these may be missing in those species which live in perpetual darkness, such as in caves. In addition, up to three simple eyes (ocelli) are often, though not invariably present. Arising from the front of the head is a pair of antennae. These vary from the simple to the complex in nature (*see* box).

MOUTHPARTS

The mouthparts are on the lower surface of the head and are perhaps more variable than any other insect organ – this relates to the way in which a particular insect feeds and also to what other uses, such as boring into wood, its mouthparts may be put. Compared with the teeth and jaw arrangement of, for example, man, the mouthparts of insects are quite complex.

THE THREE MAIN ORIENTATIONS OF INSECT HEADS

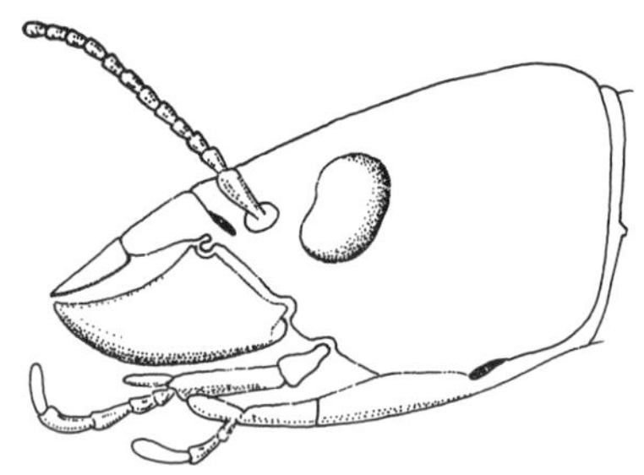

a. the prognathous arrangement, e.g. some beetles

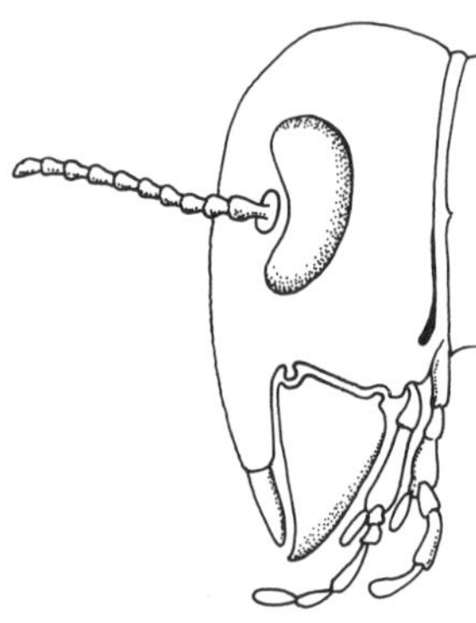

b. the hypognathous arrangement, e.g. many grasshoppers

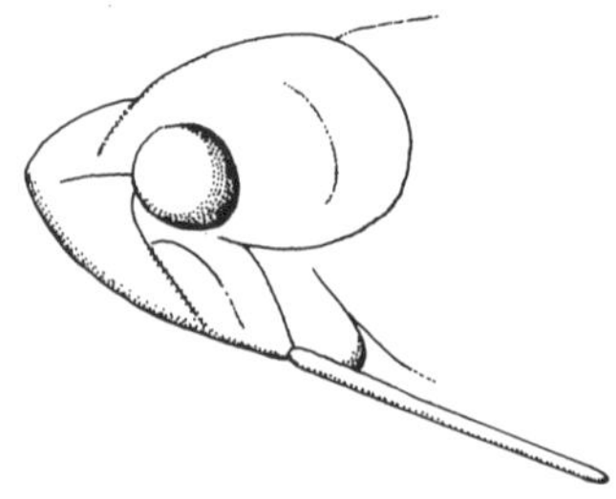

c. the opisthognathous arrangement, e.g. a cicada.

ANTENNAE

The pair of antennae articulate with the head either between the eyes or in front of them. In many instances they are simply a slim filament made up of many similar-sized segments. In higher insects, these segments may be modified in a number of different ways to produce the range of antennal types illustrated.

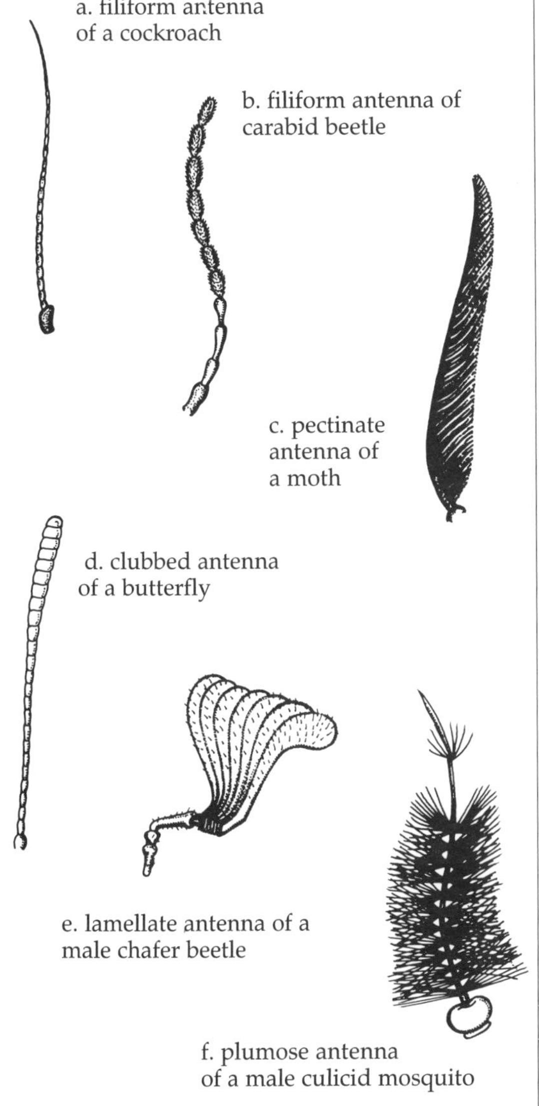

A front view of the social wasp, **Polistes carnifex**, from Mexico, feeding on the surface of a solanum fruit. Careful examination reveals the paired antennae with between and just above them two of the three ocelli visible. The large, compound eyes are obvious as are the black-fringed mandibles of the chewing mouthparts. Just discernible, peeping from behind the wasp's left mandible, is one of the palps with which it tastes the food.

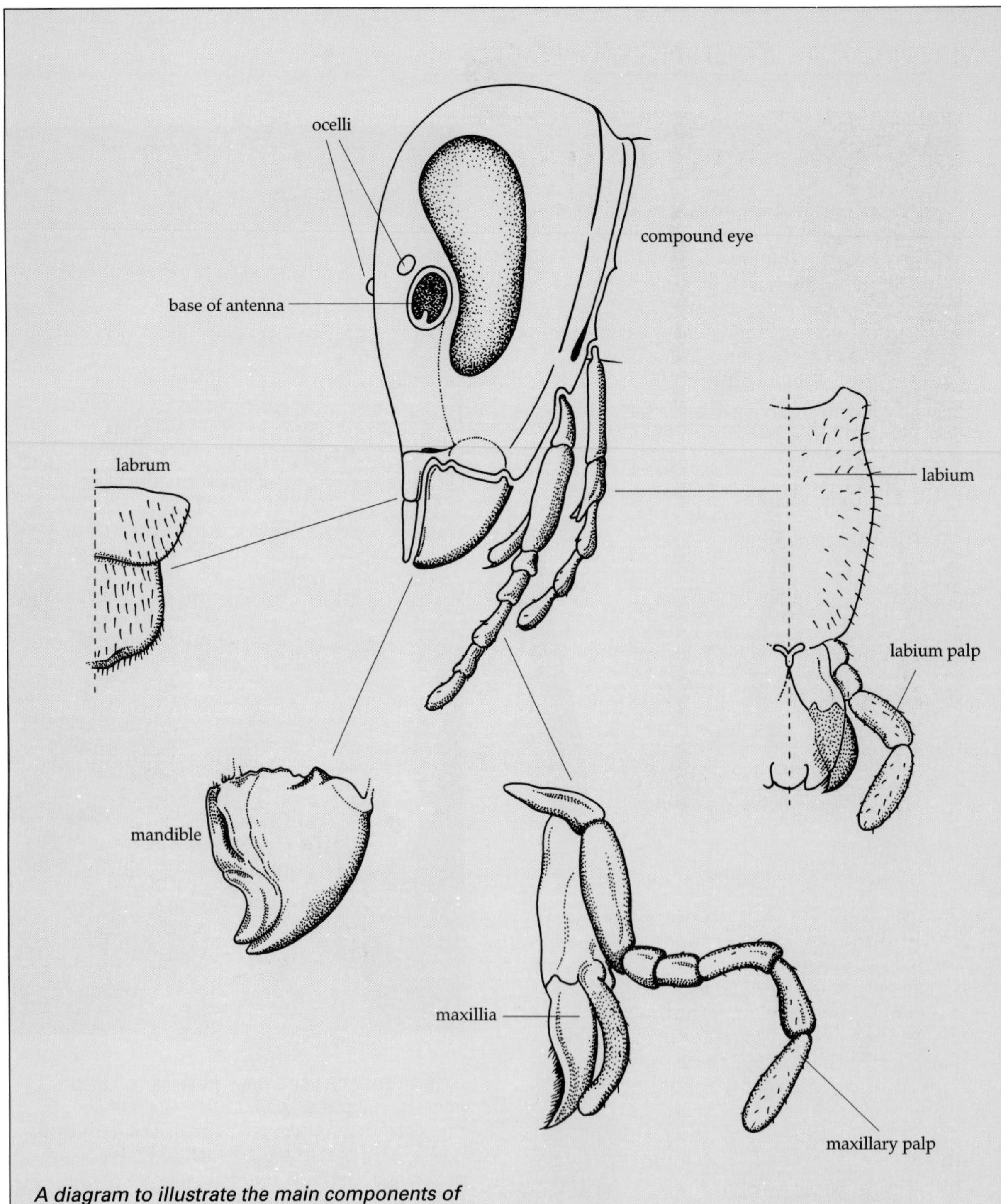

A diagram to illustrate the main components of an insect with chewing mouthparts.

BITING AND SUCKING MOUTHPARTS

The slim green proboscis is clearly visible as this ***Charaxes subornatus*** *butterfly feeds on civet dung in a Ugandan rainforest. When out of use the proboscis is tightly coiled beneath the head; unrolling to its full extent to feed, blood is pumped down through special channels.*

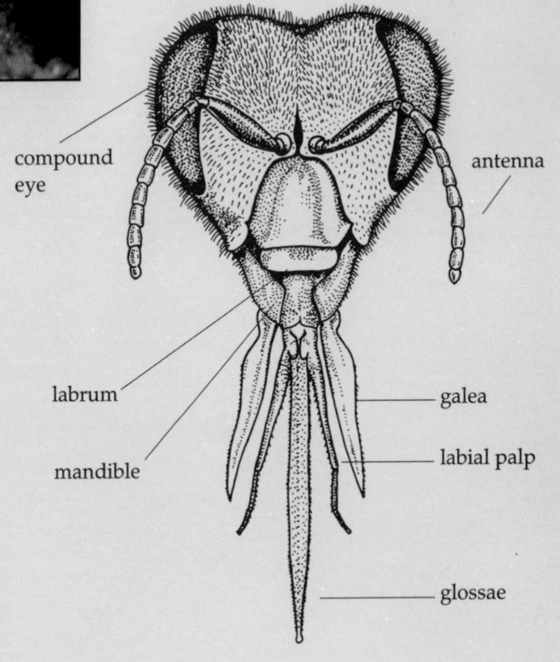

Frontal view of the head of a honey bee showing the arrangement of a set of sucking mouthparts. These have evolved from the basic chewing mouthparts of the type illustrated in the diagram opposite; other types of sucking mouthparts, for example those of mosquitoes, have evolved differently. In the bee the mandibles are still present and are used for manipulating nest-building and comb-making materials. The galea of the maxillae and the labial palps have lengthened and form a tube around the glossae, which are extensions of the labium, the lower lip. Together these lap up the nectar and then direct it into the mouth.

Since chewing mouthparts evolved first, the diagram of the generalized insect bears such mouthparts, but all other types of feeding mechanism have evolved from the basic chewing type, even the coiled proboscis of butterflies and moths. The comparative structures of mouthparts employed in different feeding methods will be considered later.

What might be termed the 'upper lip' of the insect is the foremost part of the head capsule, the labrum, which is usually hinged onto the front of the head. Behind or below this, depending upon the way the head is held, is a pair of moveable mandibles followed by a pair of similarly moveable maxillae, both of which are used in chewing. Behind or below the maxillae is the labium or lower lip. Attached to the outer surface of each maxilla and to either side of the labium there is an appendage called a palp, which is rich in sensory structures. An example of how the basic chewing mouthparts have evolved to encompass an alternative feeding method is illustrated in the panel (previous page).

The European firebug, **Pyrrhocoris apterus**, is one of a number of bugs which specialize in feeding on seeds. It forces its proboscis (the slim, black tube passing down from the front of the head and through the lower part of the proboscis sheath), into the seed and then injects digestive enzymes into it. The digested seed contents are then pumped up into the insect's gut. Compare these sucking mouthparts with the biting mouthparts of the social wasp (page 13).

THE THORAX

Whereas segmentation of the head, which attaches to the thorax by means of a distinct neck region, is not readily visible from the exterior it is fairly easy to see that, on most adult insects, the thorax consists of three segments, since each bears a pair of legs. The three segments are, from front to back, the prothorax, the mesothorax and the metathorax. A complex series of plates on the top, bottom and sides of each thoracic segment surrounds the thoracic cavity. They also form the attachments for the legs and wings. Each of the segments bears a pair of jointed legs but the wings are on the meso- and metathorax only. The upper plate of the prothorax may be considerably enlarged to form a shield called the pronotum.

THE ABDOMEN

The part of the body, which clearly shows that insects are segmented organisms, is the abdomen. The maximum number of abdominal segments in insects is eleven. In the majority of insects segments I to VII are most readily visible and are structurally similar, though segment I tends to be reduced in size. In the Hymenoptera (wasps, bees and ants) this first segment is fused with the third thoracic segment to produce the familiar 'waist' of this group. Like the thorax, the abdominal segments are made up of a number of distinct plates with a thinner membrane between each of the segments permitting some flexibility of the abdomen.

Abdominal segments VIII onwards are modified in various ways to aid reproduction. External reproductive structures may be present on segments VIII and IX of female insects and segment IX of males. Segment XI carries a pair of cerci (types of feelers) in the lower insect orders but is absent in the more advanced insects.

There are a number of openings in the wall of the abdomen. Spiracles, the external openings of an insect's respiratory system, are found on segments I to VIII, though these are reduced in number in some insects. The reproductive openings are found in the membranes between segments VIII and IX and between segments IX and X. The anus opens through the terminal abdominal segment.

LOCOMOTORY STRUCTURES

As a rule, insects move around using their legs or wings though many soft-bodied larvae move in a worm-like fashion using hydrostatic forces. Wings are clearly used for flying and gliding but as well as walking, legs may also be used to swim, dig and jump. In nymphalid butterflies the front pair of legs is often reduced to a stump so that they appear to have just two pairs.

LEGS

Walking in insects is quite simple and based upon the premise that a tripod is very stable. The insect simply raises, for example, the left forelegs and hindlegs and the right midleg. Then, poised on the tripod formed by the other three legs, it pushes the body forwards on these, at the same time moving the raised legs forwards. At the point where the body reaches its maximum forwards progression, it then drops the raised legs down, raises the other three and repeats the process. Running just involves doing the whole thing faster. Assistance to prevent slipping on smooth surfaces is provided by claws and suckers on the insect's feet.

Insects, such as grasshoppers and flea beetles, can jump very long distances in relation to their body size. Flea beetles only 2.5mm in length can, for example, leap heights and distances of between 500 and 600mm, many times their own body length and many fleas are able to leap equivalent distances. To jump, flea beetles, fleas and grasshoppers have enlarged and highly muscular hind femurs. They contract the large muscles in the femurs thus storing energy by flexing the tibia, or in the case of fleas in an elastic resilin pad. Sudden release of the tensions set up catapults the insect upwards and forwards.

Swimming in adult insects such as water beetles and water bugs usually involves using the hind legs as paddles. An increase in the leg surface area to form these paddles has usually been achieved by

VARIATIONS IN INSECT LEGS

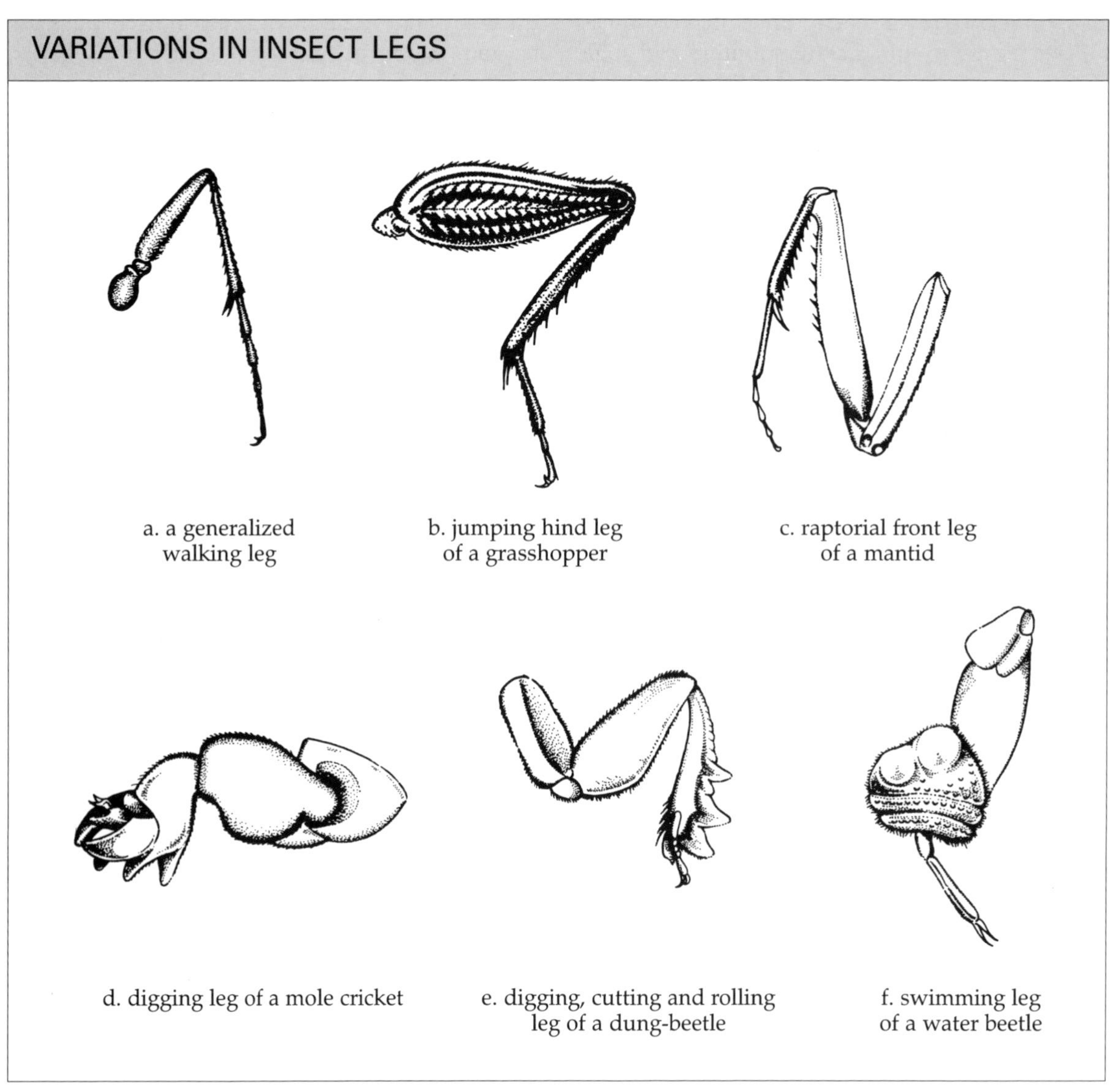

a. a generalized walking leg

b. jumping hind leg of a grasshopper

c. raptorial front leg of a mantid

d. digging leg of a mole cricket

e. digging, cutting and rolling leg of a dung-beetle

f. swimming leg of a water beetle

flattening of the legs and inclusion of rows of hairs or spines along the margins. During the driving thrust of the legs these hairs or spines are fanned out whilst during the recovery stroke they are folded up against the leg.

WINGS

The evolution of flight in insects allowed them to invade new habitats, made access to new food sources easier and also permitted them to escape from predators. The flight muscles which drive the wings may be one of two types. In more primitive insects, such as dragonflies and cockroaches, direct flight muscles are present running between the ventral thoracic plates and the wing base. Muscles on the inside of the pivot contract to produce an upward stroke of the wing, whilst those on the outside produce a downward stroke on contraction.

Higher insects retain their direct flight muscles which they use to fine tune the position and angle of the wing whilst using indirect flight muscles to power the wings. One set of these muscles runs from the roof (tergum) to the floor (sternum) of the thorax whilst the other runs from front to back. Contraction of the vertically running muscles pulls the tergum, to which the very bases of the wing are attached, downwards. This has the affect of sweeping the main part of the wing upwards about its pivotal point, at the same time somewhat lengthening the thorax which, being fluid filled, cannot change

THE WING ARRANGEMENT OF SOME INSECT ORDERS

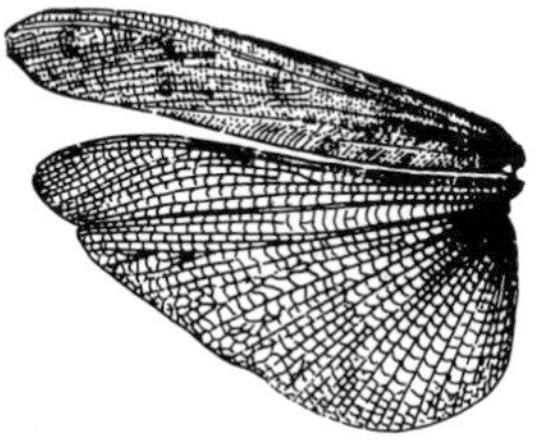

a. a grasshopper with leathery forewings protecting membranous hindwings

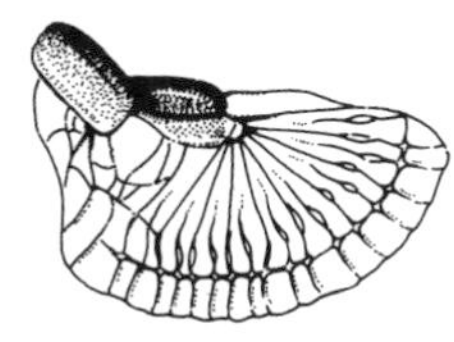

b. an earwig with short, leathery forewings which protect specially folded, membranous hindwings

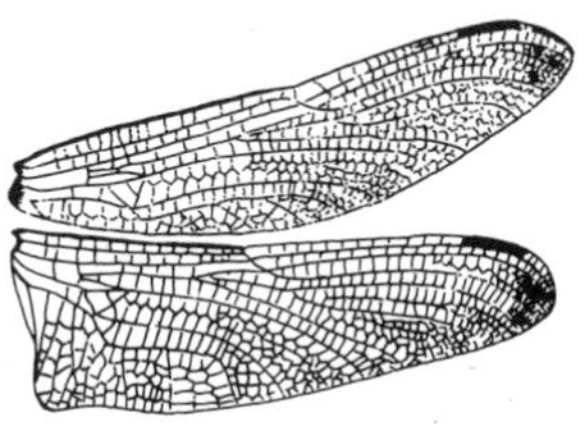

c. dragonfly with both pairs of wings membranous

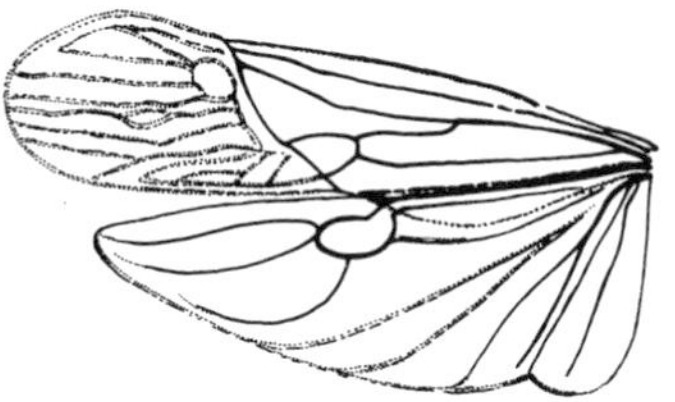

d. heteropteran bug with forewings leathery except for apical area covering membranous hindwings

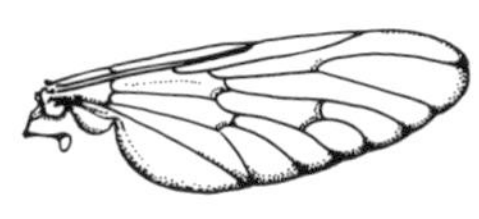

e. fly with membranous forewings; hindwings modified as halteres

f. butterfly, both pairs of wings membranous, covered in feather-like scales

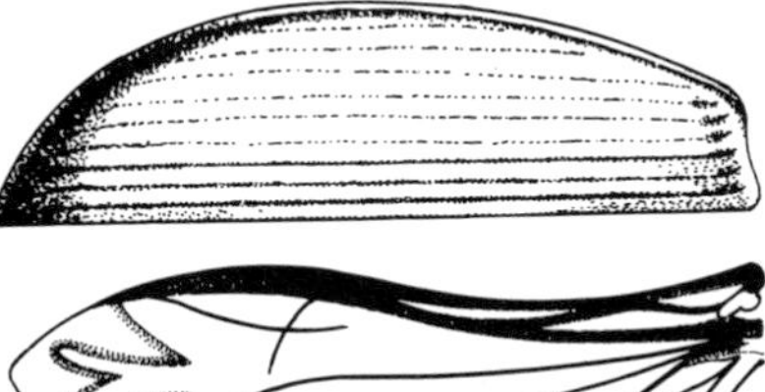

g. beetle, with heavily sclerotized forewings, the elytra, covering the membranous hindwings

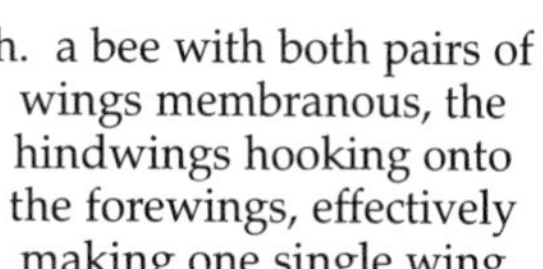

h. a bee with both pairs of wings membranous, the hindwings hooking onto the forewings, effectively making one single wing

A longhorn beetle, **Tragocephala crassicornis**, in the act of taking off from a thorny bush in a tropical dry forest in Madagascar. As in all beetles that can fly, the elytra, black and white here, are held out to the side while the transparent hindwings are actually used for flying.

its volume. The downbeat of the wings is produced when the longitudinal muscles shorten the thorax, forcing the tergum back up again. This causes the wing base to be pushed upwards sweeping the main part of the wing downwards about the pivotal point. At the end of each muscle contraction wing movement in the opposite direction is then facilitated by energy stored up in the distorted thorax, which is elastic and always contracts back to its resting shape.

Most insects in active flight have a constant wing beat though grasshoppers, dragonflies and to a certain extent butterflies exhibit a good deal of gliding flight. Before taking flight, the insect may actually spring into the air using its legs before the wings come into action, this being stimulated by the loss of contact with the ground by receptors in the insect's feet. The legs also have to absorb the shock when the insect lands. During gliding, the wings are held outstretched, changes in height being achieved simply by altering the angle of attack of the wing. Compared to aircraft, which usually stall with a wing angle of attack of around 20°, the angle on an insect wing can increase to as much as 50° before stalling occurs, giving them a great degree of manoeuvrability in the air. During active flight, the wings are alternatively driven forwards and downwards and then backwards and upwards, with a simultaneous rotation of the wing about its base. Changes in angle of attack of the wings and the degree of rotation of one wing in relation to the other allow the insect to climb or turn, respectively.

The physiology of wing muscle action is very interesting. In insects with a slow wing beat (100 beats per second or less) the muscles are synchronous, that is the nervous system is able to maintain one impulse per muscle contraction. At rates above this level (to greater than 1000 beats per second) the muscles are asynchronous, which means they contract several times following each nerve stimulation. They do this because they have an inbuilt property which allows them to contract again as soon as any tension on them is released. Thus when the wing is fully up, tension on the now stretched longitudinal muscles is suddenly released and they can contract forcing the wing down. At the end of the downwards stroke, tension on the stretched vertical muscles is suddenly released and they, in turn, can contract.

THE EXOSKELETON

It was explained in Chapter 1 that the exoskeleton plays a major part in the insect success story. What is the exoskeleton made of, how is it formed and how does it function?

Covering as it does the whole outer surface of the insect the exoskeleton is also effectively its skin or integument. It is the structure and properties of the integument that we will consider initially. The integument consists of an outer cuticle and an inner layer of epidermal cells. The latter, which secretes the cuticle, is the outermost layer of living cells covering the insect body and is called the epidermis. Under the light microscope the cuticle is seen to consist of two main layers, a thin outer layer called the epicuticle and a thicker inner layer, the procuticle.

Greater magnification of the cuticle using the electron microscope reveals that these two main layers are themselves subdivided. The epicuticle may consist of up to five layers but usually consists of an inner epicuticle followed by an outer epicuticle, which in turn is covered by a superficial layer. The superficial layer has a covering of wax which plays a major part in preventing water loss from the insect's outer surface. Very high magnification of the outer surface of the cuticle reveals that it is not completely smooth. In those insects which are relatively smooth and shiny the surface is seen to consist of polygons, often hexagons, which have only slightly raised edges. This relatively smooth surface is, however, often obscured by knobs, folds or spines, which reduces light reflectivity giving a duller look to the insect.

CHEMICAL STRUCTURE OF THE CUTICLE

The two major components of the insect procuticle are the substance chitin and the cuticle proteins. Chitin is a long-chain polysaccharide consisting of repeating residues of an amino sugar, N-acetylglucosamine. These long molecules are not unlike those of cellulose, which performs the same function of support in plants as chitin does in insects. In the insect cuticle, the chitin molecules occur as crystalline rods and bundles of these are bound together by the cuticular proteins. This arrangement gives the cuticle its combined properties of high tensile strength with a high degree of flexibility.

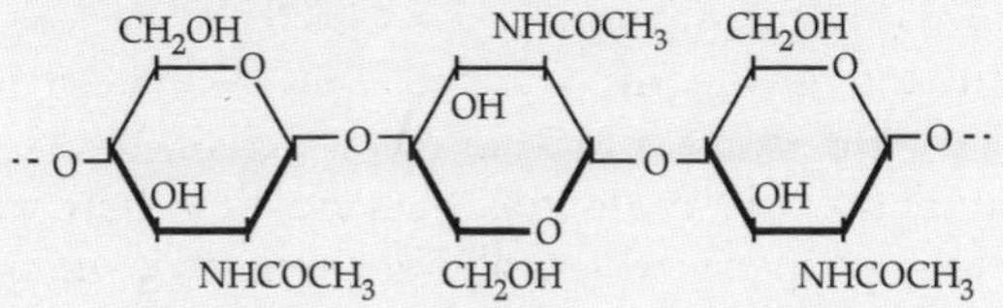

A further strengthening of the exocuticle takes place as a result of sclerotization. During this process the exocuticle proteins become irreversibly altered (in the same way that egg white becomes irreversibly changed when it is heated). They lose their solubility, become stiffer and much darker in colour. These changes are noticeable following moulting in an insect such as a cockroach, which is pale and soft when it first emerges from its old skin but it rapidly darkens and hardens off.

At the opposite extreme to highly sclerotized plates are the thin, flexible and to some degree elastic arthrodial membranes. These contain more chitin than the sclerotized areas of the body and the protein binding the chitin is not sclerotized. This combination maintains a high tensile strength whilst allowing folding and a greater degree of flexibility.

In certain areas of the cuticle may be found the protein resilin. This is somewhat like rubber in nature and is able to store up energy when it is bent, springing back into its original shape when released. This protein is found in insect wing ligaments and is also associated with the jumping of insects, such as fleas.

MOULTING IN INSECTS

Since the exoskeleton has a limited capacity to stretch during growth, it has to be renewed at intervals in order for the insect to increase in size. This renewal involves formation of a new cuticle under the old one, which is then moulted, a process referred to as ecdysis.

The first thing that happens is that the old cuticle separates from the epidermis. This creates a gap into which the epidermis secretes a fluid containing moulting enzymes. At the same time the epidermis produces a new epicuticle followed by a procuticle, which at this stage is not differentiated into the exocuticle and endocuticle. Meanwhile the moulting fluid digests the old endocuticle so that some of its constituents can be recycled, since once digestion is complete, the fluid is reabsorbed back into the insect. Once the new cuticle is complete, and the fluid has been reabsorbed, the old cuticle splits open and the insect emerges from it. The cast skin, or exuviae, consists of undigestible chitin and other compounds and is usually discarded, though some insects, for example some nymphal katydids, eat the old skin. The new cuticle is still soft and folded at this stage so the insect takes in air, or water if it is aquatic. This increases the pressure inside and expands the body and thus the cuticle to its new size. The procuticle now differentiates into exocuticle and endocuticle and the latter can continue to increase in thickness for some time after moulting is complete. The exocuticle in the meantime stiffens up and may become sclerotized.

Ecdysis in the four-spotted chaser dragonfly, *Libellula quadrimaculata*

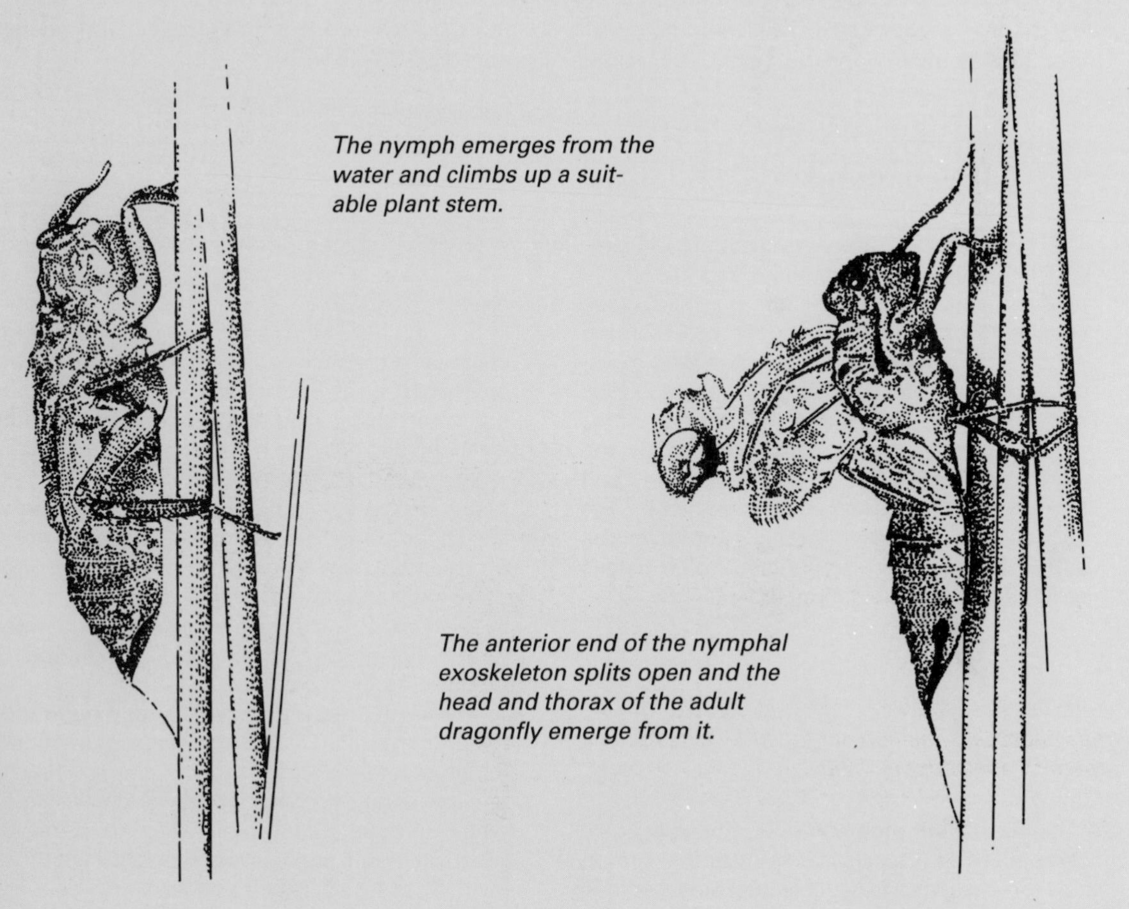

The nymph emerges from the water and climbs up a suitable plant stem.

The anterior end of the nymphal exoskeleton splits open and the head and thorax of the adult dragonfly emerge from it.

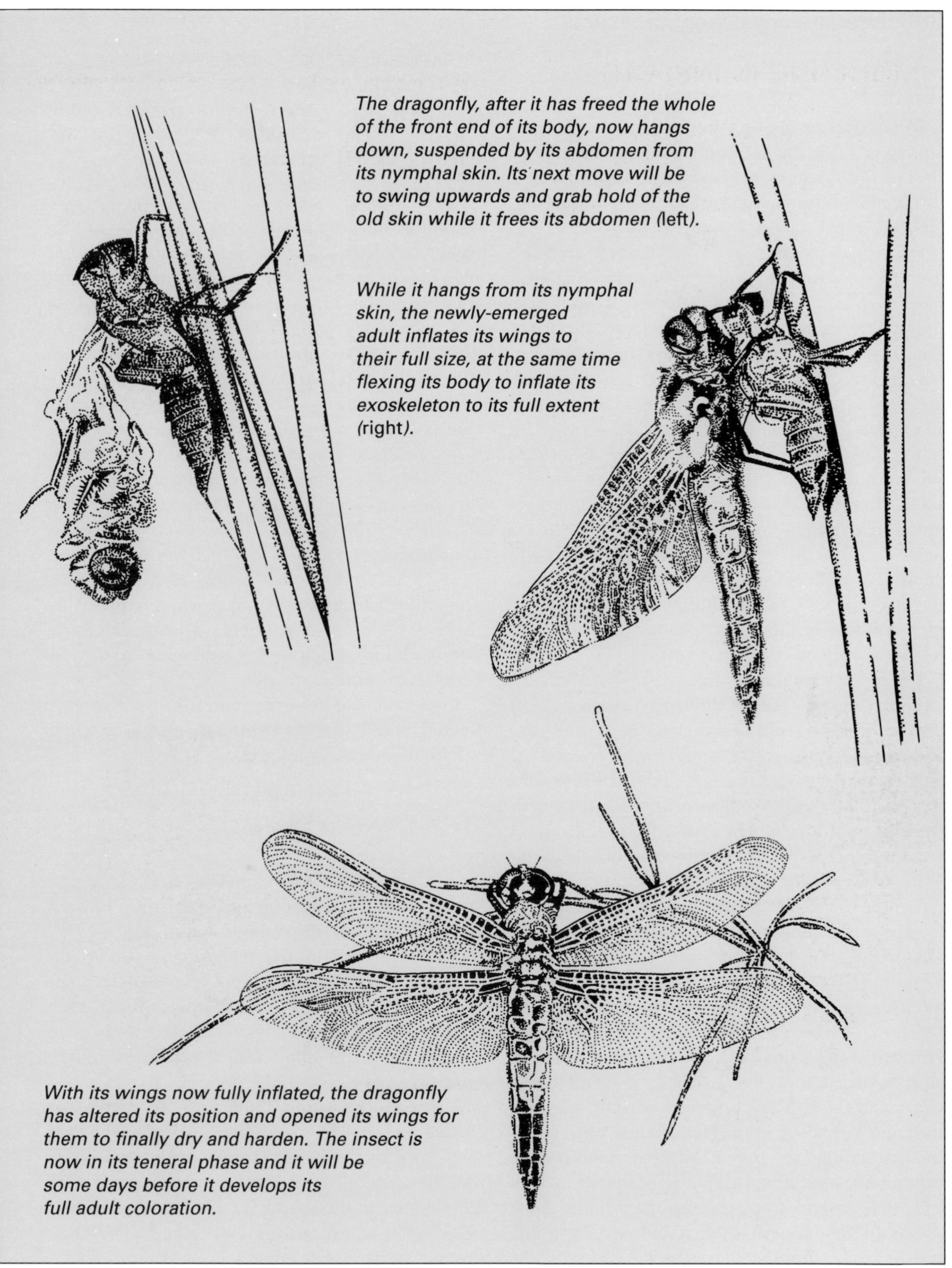

*The dragonfly, after it has freed the whole of the front end of its body, now hangs down, suspended by its abdomen from its nymphal skin. Its next move will be to swing upwards and grab hold of the old skin while it frees its abdomen (*left*).*

*While it hangs from its nymphal skin, the newly-emerged adult inflates its wings to their full size, at the same time flexing its body to inflate its exoskeleton to its full extent (*right*).*

With its wings now fully inflated, the dragonfly has altered its position and opened its wings for them to finally dry and harden. The insect is now in its teneral phase and it will be some days before it develops its full adult coloration.

In contrast to the very thin epicuticle, the procuticle is much thicker and constitutes the major part of the integument. Normally this consists of two layers, a thinner, outer exocuticle, which is pigmented, and a thicker, inner endocuticle.

Pore canals pass from the epithelium through the procuticle. Where the procuticle and epicuticle meet the pore canals form junctions with wax canals, the latter terminating at or near the outer surface of the cuticle. It is thought that wax and lipids (fats and oils) produced by the epidermis are transported to the outer layers of the cuticle through the pore and wax canals.

There is evidence that in some insects at least the amount of wax produced varies with the seasons. The beetle *Eleodes armata*, for example produces more wax in the summer, when water loss is likely to be at its greatest, than they do in the damper months of winter. Furthermore, if they are physically removed from winter conditions to summer conditions their wax production increases from the winter to the summer level.

The insect integument shows a great deal of variation in thickness and flexibility. The plates which form the head capsule, the thoracic and the abdominal segments, are usually highly sclerotized as are the wing cases in insects such as beetles, and the mandibles and maxillae of those insects with chewing mouthparts. The arthrodial membrane is much thinner and more flexible and occurs at the joints between the plates. This soft cuticle covers the whole of the thorax and abdomen in many larvae and allows them to grow by limited expansion. The greatest degree of extension of the arthrodial membrane is to be seen in insects such as queen termites and fully engorged honeypot ants, both of which can have enormously stretched abdomens. This stretching is due both to a degree of elasticity in the cuticle and to the fact that it is formed into a series of folds in the unstretched state – as the abdomen stretches, so the folds flatten out. A similar unfolding of the arthrodial membranes between the abdominal segments of female locusts and other ground-laying grasshoppers allows them to extend their abdomens deep below the ground in order to lay their batches of eggs.

INTERNAL STRUCTURE OF INSECTS

As well as the more obvious soft internal organs the insect does have a limited endoskeleton. The internal parts of the skeleton are called apodemes and they are formed from intuckings of the exoskeleton, which harden off and become rigid. The endoskeleton is subdivided into the tentorium, which is the internal skeleton of the head and the endothorax, which is the internal skeleton for the thorax. The tentorium serves three main functions. Firstly, it acts as a support for the brain and the front part of the intestine; secondly, it strengthens those parts of the head capsule onto which the mouthparts hinge; and thirdly, it acts as points of attachment for the muscles which work the mouthparts. The endothorax consists of a number of apodemes which serve as sites of attachment for thoracic flight muscles. There is also an abdominal endoskeleton, which serves as attachment sites for various abdominal muscles.

Internally, the insect contains most of the usual animal organ systems (opposite), each of which will be dealt with in turn.

THE ALIMENTARY CANAL

The alimentary canal contains structures concerned with the digestion and absorption of food. It is subdivided into three main regions, the foregut followed by the midgut and finally, the hindgut. During development of the insect from the egg, the foregut and hindgut form from intuckings of the outer body wall and, as a consequence, they are lined with a chitinous cuticle, which is not present in the midgut.

A true mouth is found only in those insects with cutting and chewing mouthparts, where it is the cavity formed by the labrum above, the labium below, and the mandibles and maxillae on either side. Usually situated in the thorax is a pair of labial glands, more better known as the salivary glands. The single ducts from each gland then join together to form a

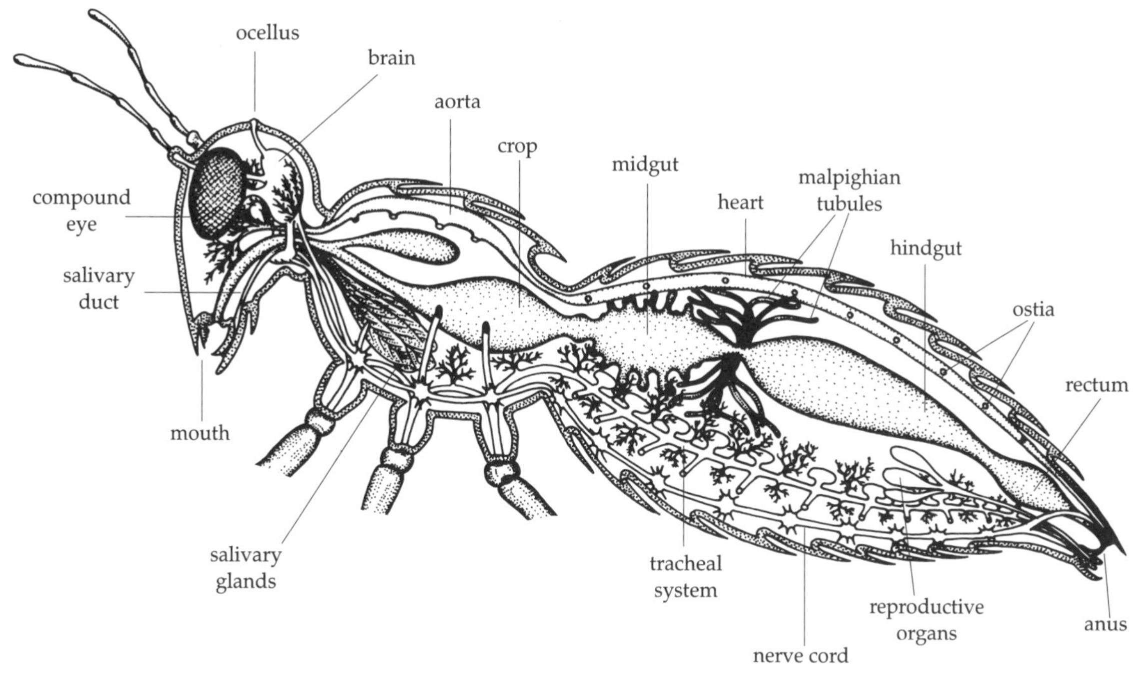

The internal structure of an insect.

salivary duct which empties into the mouth through the labium. The salivary glands are found in most insects group and they vary a great deal in both structure and function. Whereas in some insects, saliva is associated with digestion, particularly of starch, in others it may act as a poison or an anticoagulant. In the Lepidoptera (moths and butterflies) and the Trichoptera (caddis-flies), the larval salivary glands function as silk glands and their saliva-producing role is taken over by the paired mandibular glands.

The mouth leads into the pharynx, which is best developed in insects with sucking mouthparts, where it functions as a pump. The pharynx is followed by a variable length oesophagus, which passes back from the head into the thorax. In many insects, the hind end of the oesophagus is dilated to form a crop which serves as a food reservoir. In some sucking insects, a side tube leads from the oesophagus into a greatly enlarged crop, which is known as the food reservoir.

The gizzard, which follows the crop, reaches its greatest development in chewing insects which feed upon hard foods. In these insects, its chitinous lining is covered in tooth-like processes and its musculature is well developed. This arrangements assists in the breaking up of the tough food on which they feed (compare this with the action of the bird gizzard). In sucking insects there may be a minimal development of the gizzard.

The gizzard is the final section of the foregut and it is followed by midgut which may also be referred to as the stomach. Whereas the main function of the foregut is to store and prepare the food, the functions of the midgut are to digest and absorb it. It is lined with a layer of cells which both produce digestive enzymes and absorb the food once digestion has taken place. Leading off from the anterior end of the midgut, at least in most insects, are a number of sacs called gastric caeca, which considerably increase the surface area available for digestion and absorption.

The posterior end of the midgut and the beginning of the hindgut is marked by the point at which the Malpighian tubules (*see* the excretory system on page 28) enter the gut. The first part of the hindgut is the ileum, which is followed by the colon and finally the rectum, whose contents pass to the outside via the anus.

THE PHYSIOLOGY OF DIGESTION

As might be expected with such a variable group of organisms there is not a single, straightforward digestion pattern in the insects and, what is more, the processes of digestion in a larva may change radically when it becomes an adult, this usually being due to a major change in the diet. Thus only an outline of the different systems employed will be given here.

Firstly, a number of insects carry out most of their digestion outside the body, by pumping digestive enzymes from the midgut either into (seed-feeding and predatory bugs) or onto (predatory beetles) their food. The digested food remains are then sucked into the gut where any final digestion, and then absorption, of the food takes place. A number of insects, including some beetles, grasshoppers and cockroaches have large crops where food is partially digested by enzymes passing forward from the midgut.

One particular food component, which forms a major proportion of the diet of herbivorous insects is cellulose, the skeletal material of plant cells. This requires the presence of the enzyme cellulase for its initial digestion. A few insects are able to produce the enzyme themselves but the majority either require the assistance of symbiotic (mutually beneficial) micro-organisms, which live within the insect, or they ingest fungal cellulase with the food.

THE CIRCULATORY SYSTEM

Cut an insect and it bleeds, more often than not to death. The reason is that insects have an open circulatory system in contrast to the closed circulatory system found in vertebrates. Instead of flowing solely within vessels, insect haemolymph (blood) flows through various compartments within the body, bathing the organs to which it flows. The figure on the previous page shows that there is in fact one main vessel, the dorsal vessel, which is effectively the heart, with a forward extension, the aorta passing to the head. In some insects, the aorta, just ends in the head near the brain and haemolymph gushes out of it, or it may divide into two or more arteries which may subdivide into smaller vessels.

The heart shows varying degrees of compartmentalization. Thus in more primitive insects the number of chambers may be as many as the maximum number of abdominal segments. There is, however, a tendency for the number of chambers to decrease with an increase in evolutionary advance so that in the housefly, one of the most advanced insects, the heart consists of just three chambers. Between successive chambers is a ventricular valve, which prevents backflow into the previous chamber while contraction is forcing haemolymph into the chamber in front, or into the aorta. On either side of each heart chamber is a non-return valve, the ostium, through which haemolymph enters from the pericardial sinus. The latter is a compartment running the length of the insect, its roof formed by the dorsal integument and its floor by the dorsal diaphragm lying above the gut. Running along the body, below the gut and above the ventral nerve cord, is a second membrane, the ventral diaphragm. A second compartment, the perineural sinus, occupies the space between the ventral integument and the ventral diaphragm. The compartment between the dorsal and ventral diaphragms is the visceral sinus in which are situated the main body organs. The presence of various other membranes throughout the system ensures that there is in fact a directional flow of haemolymph around the insect body.

Unlike that of the vertebrates, the haemolymph system of insects has little to do with oxygen transport, though it does assume all of the other roles of vertebrate blood, that is it transports food and waste products and carries clotting compounds to seal up small wounds. The haemolymph also contains cells (haemocytes) which behave in much the same way as human white blood cells by attacking invading micro-organisms.

THE PHYSIOLOGY OF THE CIRCULATORY SYSTEM

The heart is the main structure pumping haemolymph throughout the insect body, though some insects possess accessory pulsatile organs, in for example the thorax, the legs or the base of the antennae, which assist the heart. The heart beat is maintained by muscles in its wall, the beat being rhythmic in nature, that chamber nearest the tail contracting first and a wave of contractions then passing forwards through each succeeding chamber. Haemolymph enters the heart chambers through the ostia from the pericardial sinus. The haemolymph is first delivered to the head from where it then passes into the appendages and, when present, into the perineural sinus. From here contractions of the ventral diaphragm force the haemolymph backwards and laterally into the perivisceral sinus. Any flow through the wings is through veins but septa (walls) divide the tubes which form the other appendages into two separate halves. Haemolymph flows along one side to the tip of the appendage and then back to the body along the opposite side. The haemolymph then flows over the organs in the visceral sinus before making its way back into the pericardial sinus to repeat the circulation.

THE RESPIRATORY SYSTEM

The system concerned with breathing and gas exchange in insects bears no resemblance to the more familiar lungs of the vertebrates. Instead, insects have a complex tracheal system of tubes (tracheae) which carry oxygen from the external atmosphere directly to the various body organs. The tracheae are quite elastic and are lined with a layer of chitin which is continuous with the cuticle covering the outside of the insect. This chitin layer has raised ridges arranged either in continuous (with occasional breaks), spirals or discrete rings. This helps to keep the tracheae open whilst allowing them a great deal of flexibility. Outside the chitin tube, the trachea is covered with a layer of epithelial cells.

The tracheal system opens to the external atmosphere through the spiracles. Each spiracle consists of an opening surrounded by a sclerotized plate leading into a chamber, the atrium. There is usually also present a closing apparatus, operated by a set of muscles. The spiracles open into large tracheal trunks running along the body and in flying insects these may open into large air sacs. These are very delicate and tend to lack the supporting chitin rings found in the tracheae. Smaller tracheae lead from the main trunks to the various organs. Here the tracheae subdivide and eventually become the tracheoles, which are small enough to permeate between the cells to which they are carrying oxygen.

THE PHYSIOLOGY OF GAS EXCHANGE

Transport of oxygen from the external atmosphere to the tissues through the tracheal system is by simple diffusion alone, though this may, in some insects, be through the pumping action of the

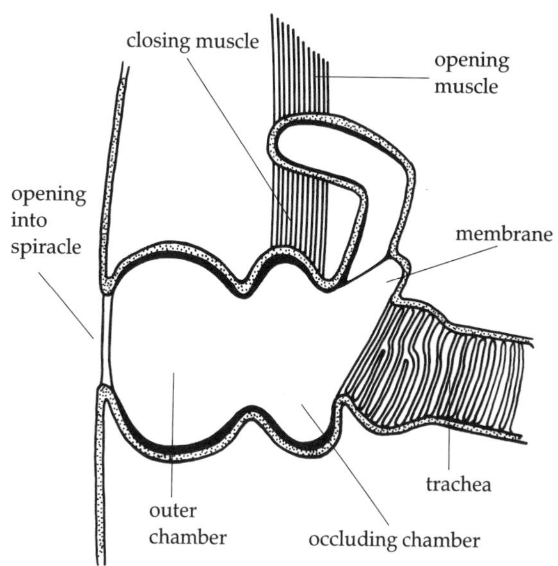

Ant abdominal spiracle. The spiracle can be opened and closed by action of the opening and closing muscles, which alter the shape of the occluding chamber. The spiracle is shown open but contraction of the closing muscle pulls the membrane across the trachea, effectively shutting it off from the outside air.

A close-up shot of the side of a notodontid moth caterpillar in rainforest in New Guinea clearly shows the single spiracle on each abdominal segment, through which air enters and leaves the tracheal system. The spiracles are the oval, red plates with black surrounds. Also clearly visible are the abdominal prolegs characteristic of lepidopteran larvae.

thoracic and abdominal muscles. By opening the thoracic spiracles during inspiration, keeping the abdominal spiracles shut, and opening the abdominal spiracles during expiration, with the thoracic spiracles closed, air is pumped along the tracheal trunks. This effectively shortens the distance that the oxygen has to travel during its journey to the tissues. Carbon dioxide diffuses out in the opposite direction, though some is known to diffuse directly out through the body surface. There is also an accompanying and unavoidable loss of water at the same time; thus, during periods of inactivity, insects close their spiracles to prevent excess dehydration.

THE EXCRETORY SYSTEM

Excretion of waste products, especially compounds containing nitrogen derived from protein breakdown, involves the Malpighian tubules and indirectly the hindgut. Nitrogenous excretion is unavoidably tied up with the control of water loss and body fluid concentrations, which thus explains the role of the hindgut, the site of these activities. The Malpighian tubules lie in the body cavity surrounded by haemolymph. Waste products are pumped from the haemolymph into the tubules through the tubule walls. At the same time, water, salts, sugars and other substances passively diffuse in the same direction. The sugars are then actively reabsorbed by the Malpighian tubules, but the remaining fluid passes into the rectum where most of the salts and water are reabsorbed, leaving a concentrate of waste products to pass out of the body through the anus. The main excretory waste product in insects is uric acid (or urates), which can be removed from the body with the minimum of water. Aquatic insects do not have a problem with water

retention so that they are able to excrete ammonia, highly toxic when concentrated, in dilute solution.

NERVOUS SYSTEM

The nervous system as a whole is subdivided into a central nervous system and a peripheral nervous system.

CENTRAL NERVOUS SYSTEM

Masses of intermediate and motor neuron cell bodies form a structure called a ganglion. In the basic insect design, there is one pair of ganglia in each segment and these are interconnected to form the central nervous system. The first three pairs of ganglia coalesce to form the brain, which lies in the top of the head. The longitudinal nerve cord then passes round either side of the head to the suboesophageal ganglion which is formed from a joining of the fourth to sixth pairs of head ganglia. The basic arrangement of the three pairs of thoracic ganglia and the eight pairs of abdominal ganglia tends to be the exception rather than the rule. In many insects, the apparent numbers of ganglia are reduced as a result of the fusing together of the ganglia in adjacent body segments. Thus some insects have three thoracic but only six abdominal ganglia, in some all the thoracic and abdominal ganglia are fused together, and in the most extreme situation, these are fused with the suboesophageal ganglion to produce a single structure, a veritable 'superganglion'.

PERIPHERAL NERVOUS SYSTEM

This is made up of bundles of nerve fibres of the sensory and motor neurons, passing from the sense organs to the central nervous system and from there to the various body muscles. The visceral (or sympathetic) nervous system is an involuntary system arising from the central nervous system. It controls those parts of the body over which the insect can exercise no active control, that is the gut, the heart, the breathing system, etc.

REPRODUCTIVE SYSTEM

The male and female reproductive structures in insects show an amazing array of different variations on a basic theme. Each male testis is made up of a number of lobes in which the spermatozoa are produced. These then pass down the vas deferens to the seminal vesicle, where they are stored. The sperm then pass to the female, either as semen or as a sperm package, the spermatophore, via the ejaculatory duct. In some insects the spermatophore consists of a sperm-containing ampulla portion and a spermatophylax, which is rich in protein and which is discussed in detail later in the chapter. The role of the accessory glands is to secrete nutrients to maintain the sperm during their passage into the female insect and to produce the spermatophore when this is present.

MALE AND FEMALE REPRODUCTIVE SYSTEMS

In the female insect, there is a pair of ovaries each of which is composed of a number of egg-producing ovarioles. From each ovary runs a lateral oviduct and the two unite to form a common oviduct down which the eggs will eventually pass into the genital chamber (or vagina), which opens to the outside through the vulva. The spermatheca stores sperm from the male after mating has taken place and before the eggs are fertilized. A duct leads from the spermatheca into the vagina. Secretions to nourish the sperm stored it the spermatheca is produced either by its own spermathecal gland or by cells in the spermatheca itself. Also opening into the genital chamber are the ducts of the accessory glands, which in most insects produce secretions to cover and protect the fertilized eggs or to fasten them to the surface on which they are laid. In those Hymenoptera which sting the accessory glands have taken over the role of poison production. In tsetse flies, the anterior end of the genital chamber is expanded to form a uterus, in which the larva undergoes its development. In this insect, the larva feeds on a milky secretion produced from the accessory glands.

THE PHYSIOLOGY OF SEX

In the primitive wingless insects, the male simply deposits a sperm package, the spermatophore, onto the substrate and the female then walks over it and receives it into her genital opening. In all higher insects, copulation takes place with the male introducing spermatozoa into the female via the penis. Insects have often quite complex sets of valves and claspers on the abdomen which are species specific. Their role is to keep male and female together during the sometimes lengthy sperm transfer and they also serve to prevent mating between different but perhaps closely related species. One species 'key' will not fit into the other's 'lock'. The sometimes prolonged and often complex courtship procedures, described in more detail in Chapter 4, serve to get male and female into the correct position for the interlocking of the genitalia and subsequent sperm transfer. In some members of the Heteroptera, Coleoptera, Hymenoptera and Diptera liquid semen is passed directly from male to female but in all of the rest of the insects, sperm is transferred in the spermatophore.

In a number of katydid species, the spermatophore bears an extra, non-sperm-carrying portion, the spermatophylax, which appears to perform a special function in these insects. Following mating, the female insect consumes the protein-rich spermatophylax with great gusto. Two theories have been put forward to explain evolution of the spermatophylax. In the first instance, it is believed that during the time taken for the female to consume it, all of the sperm in the spermatophore ampulla can transfer into the reproductive tract. The second possibility is that the spermatophylax is a contribution by the male to help in the production of the offspring. By providing the female with a protein meal, the female is likely to lay more or larger eggs. Research into the role of the spermatophylax points to there being some truth in both theories.

A further discussion of the reproduction system is given later in the chapter which considers life-cycles.

Two male mosquitoes, **Culiseta annulata**, are here holding onto a female with whom they both attempted a mid-air mating. Notice on the males the inflated, feathery antennae which are highly receptive to the high-pitched whine produced by the female's wing beat and by means of which they track her down.

SENSORY SYSTEMS

VISION

Although, strictly speaking, vision entails the formation of images of surrounding subjects, in a broader sense it can also encompass the organism's ability to detect the difference between light and dark. This latter ability is important to most insects in determining the difference between night and day and in the setting of their biological clock. It also allows them to determine absolute daylength, an indication of changing seasons. Thus they are able to adjust their life cycles so that they are active at a time when food is available and are in some form of resting phase, which can be egg, larva, pupa or adult, during those seasons which are unfavourable to them.

Appreciation of daylength appears to be a major function of the ocelli. A number of insects that lack any visual organs are still, however, able to respond to light, apparently due to the presence of light detectors spread over the body surface. Even insects with eyes, when these are covered, are still able to respond to light, indicating that they also have light detectors on the body surface. Aphids have been shown to actually have light-detecting cells within the brain with which they are able to adjust their method of reproduction to seasonal changes.

Visual structures possessed by adult insects are the ocelli, already mentioned and the compound eyes. Structurally the ocelli are much simpler than the compound eyes. They do not form a sharp image but they are very sensitive to both low levels of light and changes in light intensity.

The ultimate image-forming organ in the insects is the compound eye. Compound eyes are found in virtually all adult insects and also in the nymphs of the Hemimetabola. The basic unit of the compound eye is the ommatidium, which is usually made up of six or seven visual cells. The number of ommatidia varies greatly within the insects. At one extreme, in the case of some ant workers, a single ommatidium is present. At the other extreme, in some dragonflies, as many as 28,000 ommatidia are found in each eye. Whereas in the ocellus, the whole mass of visual cells is covered by a single corneal lens, in the compound eye each ommatidium has a sharply demarcated region of the cornea, the facet, directing light into it. These facets show up clearly in the photograph of the compound eye. The pigment cells surrounding

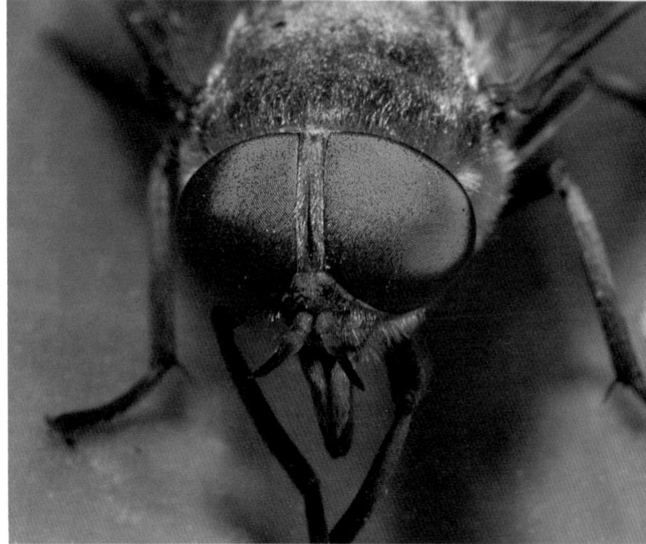

This rather sinister-looking green-eyed insect is a horsefly or deerfly of the genus **Philopomyia** from Europe. The hundreds of facets on each of the compound eyes are clearly evident as is the blunt proboscis, which delivers such a painful bite.

each ommatidium prevent light from spilling over into neighbouring ommatidia, so that each one acts as a single visual unit. The image formed by the sum of the ommatidia is not particularly sharp but absolute visual acuity is probably not of great importance to an insect. The compound eye is, however, very good at detecting movement, important for avoiding predators and in determining the insect's position relative to surrounding objects, a great asset for flying creatures.

The compound eye comes in two basic forms. The apposition eye has the basic type of ommatidia described above and works well in diurnal (day-active) insects when light intensities are high. Nocturnal insects have superposition eyes capable of working at low light intensities. In this type of eye, the ommatidium is modified so that pigment surrounding the light-sensitive cells can migrate outwards. This increases the sensitivity of the whole eye by permitting light entering one ommatidium to pass to its neighbours through its now transparent walls.

Many insects have colour vision, of special importance to those species which forage on flowers. Insect compound eyes tend to be more responsive to ultraviolet light and consequently many plants have evolved flowers which reflect attractive patterns in ultraviolet light. Insects are also able to appreciate the plane of polarized light in natural sunlight and they are able to use this in navigation.

A third visual organ, the stemma, is the only type of eye found in the larvae of holometabolous insects. In the early literature, stemmata were referred to as the 'lateral ocelli' and it is only in fairly recent times that the difference between the structure and functioning of larval stemmata and the true ocelli has become thoroughly understood. Stemmata vary in number from a single pair (one stemma on each side of the head) in, for example, sawflies, to seven pairs in Neuroptera and Megaloptera. They are, however, absent from the larvae of fleas, bees, ants and wasps.

The stemmata are actually able to form an image, though not as clearly as with a compound eye. This image is formed by the sum of the images received from all of the stemmata. Thus those larvae, such as those of tiger beetles and lacewings, which lead an active predatory existence have the greatest number and development of the stemmata.

HEARING

What precisely is meant by 'hearing' in respect to insects is not quite so clear-cut as the same phenomenon in man and other vertebrates. This is because insects are also able to detect vibrations through the surface on which they are at rest as well as through the air, and the receptors for the two processes are very similar. There are on many insects, however, structures which are analogous to the ears of vertebrates. The general name given to

A caterpillar of the drinker moth, **Philudoria potatoria**, feeding on a piece of grass. On the head capsule can be seen the shiny black, chewing jaws and a group of stemmata, special eyes found only in larval insects. The larva is holding onto the grass immediately below the head with its thoracic legs whilst further along the body it holds on with its abdominal prolegs. The dense hairs make it an unpleasant mouthful for any would-be attacker.

the structures which detect vibrations in insects is chordotonal organs. The basic functional unit of the chordotonal organ is a group of three linearly arranged cells called a scolopidium. Insect chordotonal organs contain one to many scolopidia, which usually – though not always – bridge a gap between two sections of integument. Movement of one of these in relation to the other, as a result of an incoming vibration, distorts the scolopidia and causes them to produce nerve impulses, which pass to the appropriate part of the central nervous system.

In all adult insects and in many larvae there is a particular chordotonal organ, called Johnston's organ, which is situated within the second antennal segment. Its main function is to appreciate the speed at which the insect is flying by detecting the amount that the antenna is distorted. In addition, in many midges and mosquitoes, it is used by the male to detect vibrations set up in the air by a flying female of the same species. Vibrations being transmitted through the surface on which an insect is standing, including water, are detected by the subgenual organ. These are found in the tibial segment of the legs of all insects, with the exception of the beetles and the flies, and different pairs of legs may be tuned to appreciate different frequencies.

What may be thought of as 'ears' may be situated in a number of different positions in different insect groups. They are usually referred to as tympanal organs, since they consist of a tympanum (equivalent to the eardrum in man) associated in some way with a chordotonal organ. The fact that ears may be found on the thorax of mantids and noctuid moths, the front legs of katydids and other orthoptera, the abdomen of cicadas, certain moths and other orthoptera and the wings of some moths and lacewings, points to their having evolved independently on several different occasions.

Relatively straightforward ears are found one on either side of the first abdominal segment of grasshoppers and locusts. The oval tympanum is situated in the body wall and is supported by chitinous rings. On the inner surface of each tympanum is a large concentration of scolopidia known as Müller's organ, from which runs the auditory nerve. Sound waves are picked up by the tympanum distorting the scolopidia in Müller's organ, which then fire off, sending an appropriate message to the brain. The ears of crickets and katydids are more complex than this. They are situated in the tibia of the front pair of legs and there is not one, but two, tympanic membranes open to the external atmosphere on one side and to the cavity of a tracheal branch on the other. In crickets, these tracheae are connected to the ventilating spiracles in the prothorax. In katydids, they connect to special acoustic spiracles and are independent of the main tracheal system. In both systems, sound waves enter through the spiracles and pass down the narrowing acoustic tracheae which, approximating a horn in shape, amplify the sound.

DETECTING MECHANICAL STIMULI

Apart from the detection of incoming vibrations, insects also have to be able to appreciate gravity, touch, pressure and position. The sense of touch is mediated through hair-like projections of the cuticle referred to as trichoid sensilla. Bending of the hair stimulates a sensory nerve ending which then sends a message to the central nervous system, fine tuning being achieved by having different sensilla sensitive to certain levels of stimulation.

As well as touch detection, groups of trichoid sensilla may be used to enable the insect to appreciate the position of one part of its body in relation to another. Thus they occur on joint membranes and as the joint moves, so the hairs are forced with greater or lesser pressure against an adjacent surface. Rather than firing off single nerve impulses, these sensilla fire off continuously at a rate proportional to their degree of bending. The degree of stretching of muscles is detected by stretch receptors within the muscles in much the same way as in vertebrate muscles. Finally, there are the campaniform sensilla, which detect stress during bending of the exoskeleton. They are found, for example, at the base of the wings and are thus able to respond to distortions in the basal membranes as the wings move up and down.

RESPONSE TO CHEMICAL STIMULI

Chemoreception plays an important part in the life of insects. Not only do they need to 'smell' and 'taste' food but females have to be able to detect the suitability of those things on which they are to lay their eggs and male and female insects are attracted to one another by their scents. The main insect 'nose' is the antennae, which detect wind-borne chemicals while 'taste' is not restricted to the mouth area alone but is also detected by the antennae, the proboscis and the tarsi of the legs. Even the ovipositors of female insects can 'taste' in order to find a suitable substrate in or on which to lay their eggs.

The structures which are employed to taste and smell are once again forms of sensilla. These may have a single opening to the outside (uniporous) or they may have many openings (multiporous). It is believed that the uniporous sensilla are concerned mainly with contact sensitivity, that is taste, while the multiporous forms are olfactory in function. It is thought, however, that they work in similar ways, chemicals diffusing into the sensillum where they bring about a change in the electrical polarity of a membrane, which in turn elicits a nerve impulse. As might be expected, sensitivity relates very much to the number of sensilla present. In some male moths, which can detect scents produced by females at very low concentrations and often from a long distance, a single antenna has been estimated to carry as many as 17,000 sensilla.

LIFE-CYCLES

It is a feature of all insects that the young hatch from the egg in a form which is, to a greater or lesser extent, different from that of the adult insect. The hatchling then feeds and grows, going through a series of moults before eventually becoming an adult insect. This developmental phase is referred to as metamorphosis. Three distinct life-cycle patterns are discernible in the insects as follows.

THE AMETABOLOUS LIFE-CYCLE

It is only in the two groups of primitive, wingless insects, the silverfish (Thysanura) and the bristletails (Archaeognatha) that we find this life-cycle. The offspring hatch from the egg resembling tiny adults, though they lack reproductive organs, which only develop before the final moult into the adult insect.

THE HEMIMETABOLOUS OR EXOPTERYGOTE LIFE-CYCLE

Insects which exhibit this type of life-cycle are in the so-called lower orders, that is, the Ephemeroptera (mayflies), Odonata (damselflies and dragonflies), Orthoptera (grasshoppers, crickets and katydids), Phasmatodea (stick and leaf insects), Grylloblattodea (rock crawlers), Dermaptera (earwigs), Isoptera (termites), Blattodea (cockroaches), Mantodea (mantids), Zoraptera, Plecoptera (stoneflies), Embioptera (web-spinners), Thysanoptera (thrips), Hemiptera (true bugs), Psocoptera (booklice) and Phthiraptera (lice). In these orders, the young hatch from the egg resembling miniature adults but with undeveloped reproductive organs and lacking wings. They feed and grow, passing through a number of moults. Over this period, the wings gradually

NYMPHS AND LARVAE: CONFUSING NAMES

As is so often the case with living things, there are some exceptions to the terminology of insect young. The aquatic young of the Odonata, Ephemeroptera and Plecoptera are usually referred to as a larva rather than a nymph but it is structurally and developmentally a nymph, very different in form from the larva of the higher insects in the next group. The name larva has stuck for these three groups, simply because the young shows little resemblance to the adult, but this is basically a result of modifications in the nymphal stages to suit them to life under water.

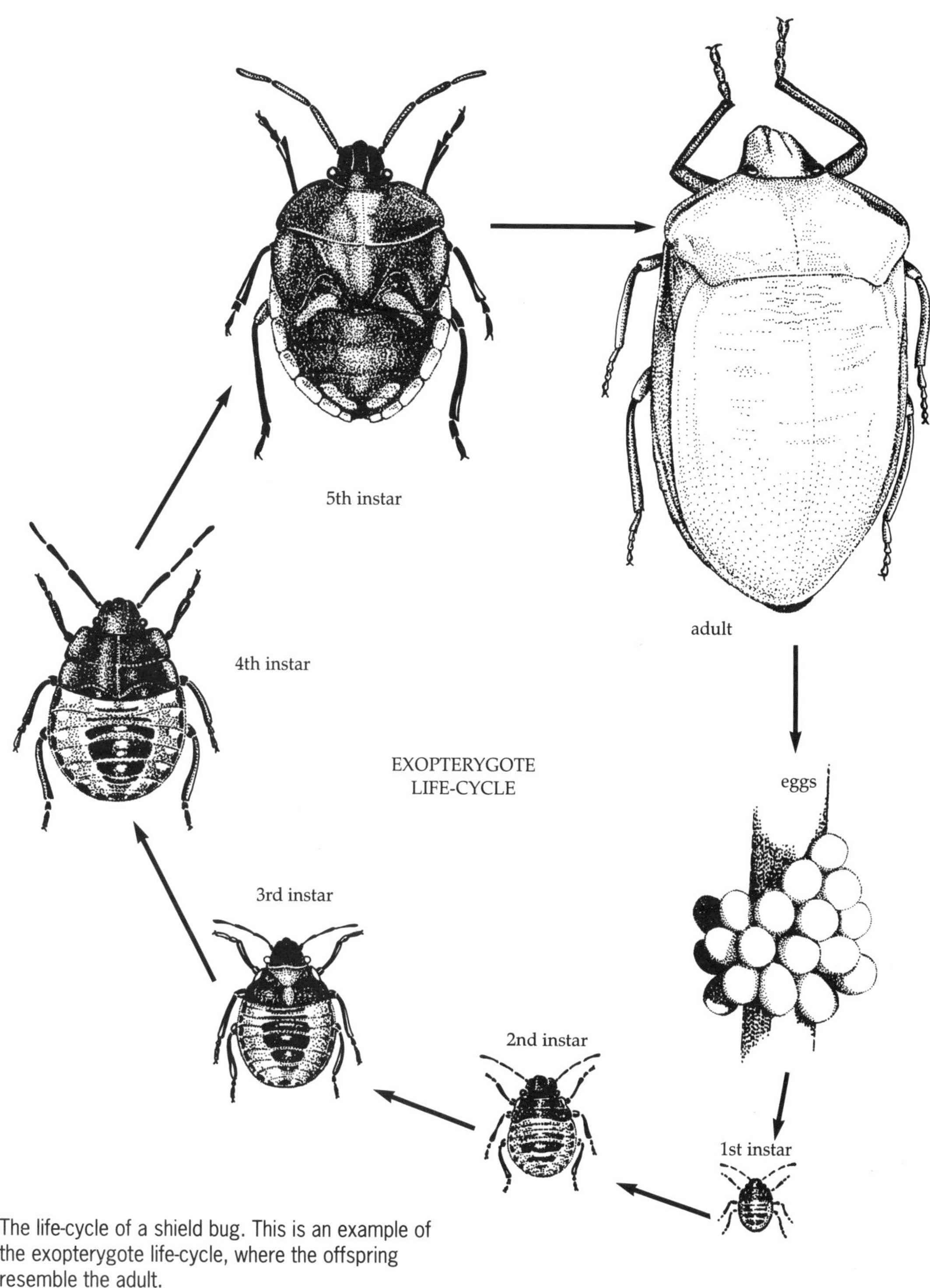

The life-cycle of a shield bug. This is an example of the exopterygote life-cycle, where the offspring resemble the adult.

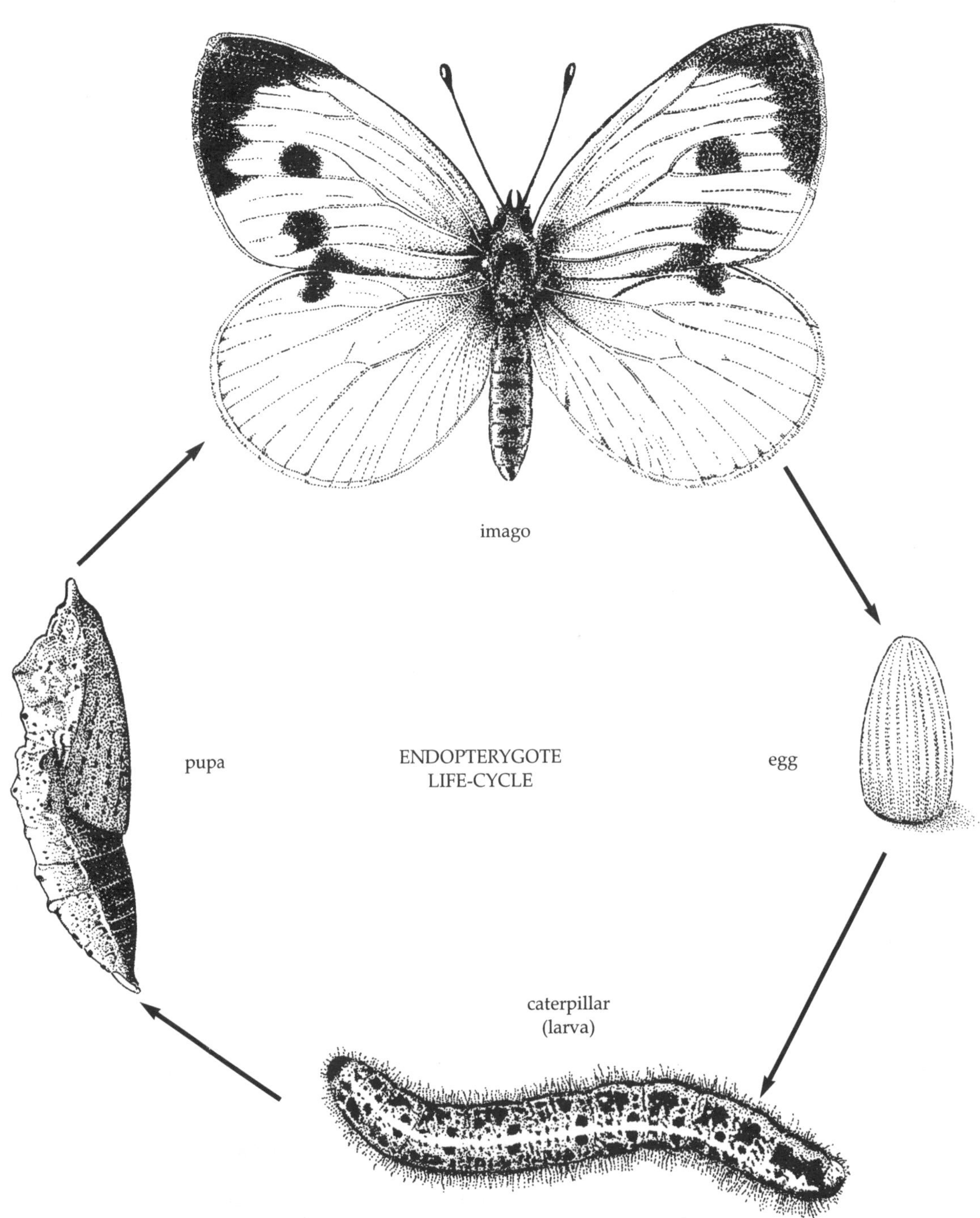

The life-cycle of the cabbage white butterfly. In the endopterygote life-cycle represented here, the offspring bear little or no resemblance to the adults.

increase in size and the reproductive organs develop. Each of the stages between moults is referred to as an instar and the young insects in each instar are usually referred to as nymphs. The final instar nymphs then moult into the adult insects, which mate, lay eggs and so the cycle is repeated.

THE HOLOMETABOLOUS OR ENDOPTERYGOTE LIFE-CYCLE

This life-cycle is characteristic of the higher insect orders, that is, the Coleoptera (beetles), Strepsiptera, Neuroptera (lacewings and ant-lions), Megaloptera (alderflies and dobsonflies), Rhaphidioptera (snakeflies), Hymenoptera (sawflies, bees, wasps and ants), Trichoptera (caddis-flies), Lepidoptera (butterflies and moths), Mecoptera (scorpionflies and hangingflies), Siphonaptera (fleas) and Diptera (flies). The young of these insects, the larva, hatches from the egg in a form which usually bears little resemblance to the adult. The larval butterfly, better known as a caterpillar, and the larval housefly, the maggot are two good examples. The larva feeds and moults at intervals, increasing in size until it finally moults to become a pupa. This often represents a resting phase in the life-cycle of the insect during which it may avoid adverse conditions such as winter cold or a tropical dry season. Inside the pupa the original larval structures are broken down to their basic components and from these the adult insect is then built up. Finally, the fully developed adult insect emerges from the pupal skin to mate, lay eggs and repeat the cycle.

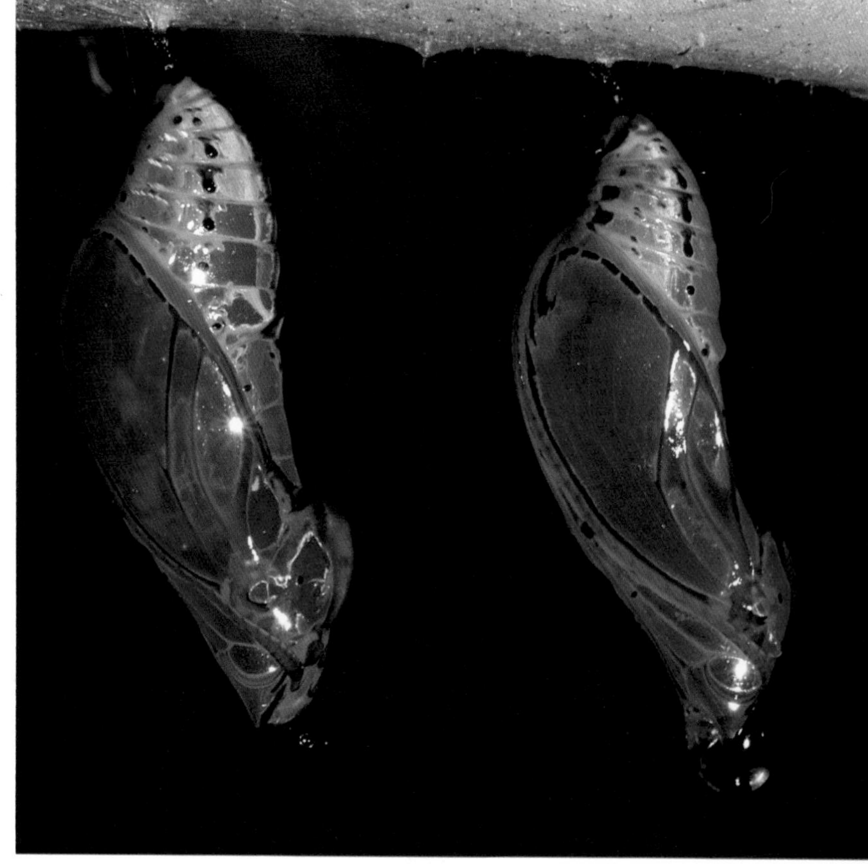

Looking for all the world like large droplets of water, hanging from a stem in Venezuelan rainforest, are these obtect, adecticous pupae of an ithomiine butterfly. Through the transparent integument the developing eyes, antennae and wings are clearly visible.

LARVAE AND PUPAE

During the development of an insect inside the egg it goes through a series of fairly clear-cut stages. Initially, in the protopod stage, there is little sign of any body segmentation, no appendages are present and the digestive, tracheal and nervous systems are in a rudimentary state. Further organ-building results in a polypod stage, in which body segments are clearly visible, appendages are developing and the internal body systems are clearly apparent. In the final oligopod stage, the thoracic appendages have increased in size and the internal body systems are fully formed. It is at the end of this stage that the so-called hemimetabolous insects emerge from the egg as a first instar nymph.

Holometabolous insects, on the other hand, show arrested development inside the egg and emerge either during the protopod or polypod stage or early on in the oligopod stage. It is thus possible to recognize four basic larval types in the holometabolous insects.

LARVAL STAGES

PROTOPOD LARVAE

There larvae are found primarily in certain parasitic wasps. The eggs contain little in the way of stored food and the larva is seemingly forced to emerge at a very early stage of development. On the other hand, since it is inside the egg or body of another insect, once it emerges it is completely surrounded by food.

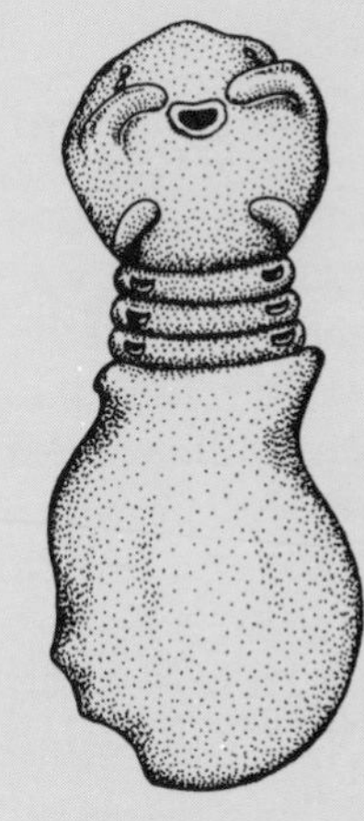

protopod larva

POLYPOD LARVAE

The larvae are familiar to most of us in the form of the moth or butterfly caterpillar, though similar larvae are also found in sawflies and scorpionflies. Polypod larvae have a well-developed head capsule with biting jaws, poorly developed thoracic legs, a variable number of abdominal legs and, with the exception of the genitals, fully developed internal organ systems. The majority of polypod larvae are vegetarians, their small legs restricting the speed at which and distance they can move, so that they remain on, or in, the vicinity of their food plant.

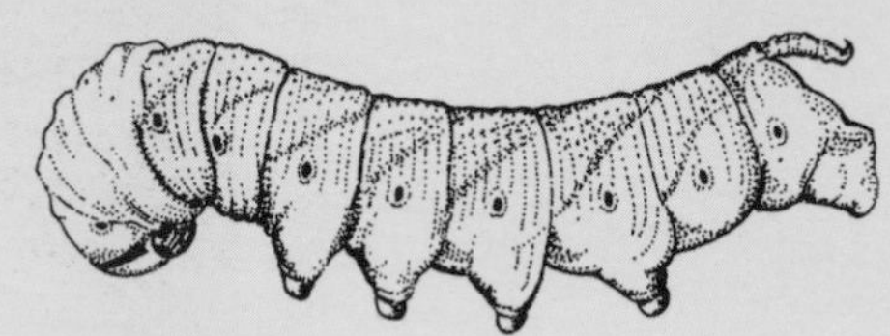

polypod larva of a moth

HEMIMETABOLOUS INSECTS

EGG

PROTOPOD LARVA → POLYPOD LARVA → OLIGOPOD LARVA

↓

1ST INSTAR NYMPH

HOLOMETABOLOUS INSECTS

EGG

PROTOPOD LARVA → POLYPOD LARVA → EARLY OLIGOPOD

PROTOPOD LARVA ↓	POLYPOD LARVA ↓	EARLY OLIGOPOD ↓
PARASITIC WASP LARVA	MOTH OR BUTTERFLY CATERPILLAR, SAWFLIES AND SCORPIONFLIES	TYPICALLY BEETLES AND NEUROPTERA

OLIGOPOD LARVAE

These larvae are, in the main, active predators. They have a well-developed head capsule with strong jaws, long thoracic limb to allow fairly rapid movement, no abdominal limbs and an armoured integument. Such larvae are typical of many beetle families and also the Neuroptera. As is so often the case with living organisms, the division between the polypod and oligopod larva is not always clear. In a number of beetle fam-

oligopod larva of a beetle

ilies, the larvae show characteristics which are a reversion towards the polypod type. Evidence for this being a reversion comes from those beetles in which the first stage larva is purely oligopod in structure but subsequent larval stages resemble polypod larvae. What seems to be a developmental reversal relates to the way in which they feed, since they are no longer active hunters but bore in wood or are parasites of other insects, so that they do not have to move any distance to find their food.

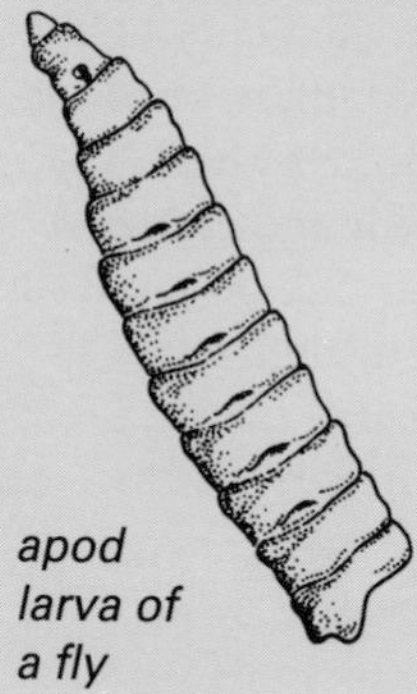
apod larva of a fly

APOD LARVAE

Most examples of these larvae, where the thoracic limbs are completely absent, are derived from the oligopod type. They are to be found in a number of beetle families, in bees, ants and wasps, fleas and the more advanced flies, where this type of larva reaches its greatest extreme in the maggot. In the last, even the head is reduced to just a pair of mouth hooks. The apod larva is, as might be expected, a modification for living in, or on, a supply of food, so limbs are not needed. In many social Hymenoptera, of course, the food is actually brought to the larvae by adult insects so they have no need to move and seek their food.

HYPERMETAMORPHOSIS

A number of insects exhibit the phenomenon of hypermetamorphosis, when two distinctly different types of larva are present in the life-cycle, representing a change in life-style from free-living to parasitic or kleptoparasitic (feeding on another's food). From the egg emerges a typical oligopod larva which then actively seeks out its host insect. Once it has found its host, it then moults to become more grub-like, resembling a polypod rather than an oligopod larva.

THE PUPA

The final larval instar moults to form the pupa where, as with the larvae, several different types are recognisable. The pupa may be covered in a cocoon spun from silk produced by the larva or the last larval skin is retained and hardens off to form an extra protective layer, the puparium, a characteristic of some higher flies.

EXARATE PUPAE

Most pupae are of this type, where the developing mouthparts, legs and wings are not held closely against the body. Exarate pupae may be dectious, that is, the mandibles are hinged and may be used to help the adult cut its way out of the cocoon. Alternatively they may be adectious, that is, the mandibles are not hinged in the pupal stage and before the adult can use them it has to escape from the pupal integument. Exarate pupae are found in the Neuroptera, Mecoptera, Trichoptera, most Coleoptera, Hymenoptera and in the primitive micropteryigidae in the Lepidoptera.

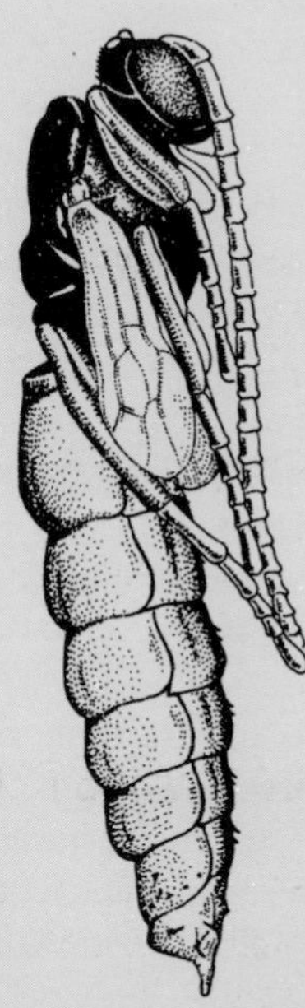
exarate pupa

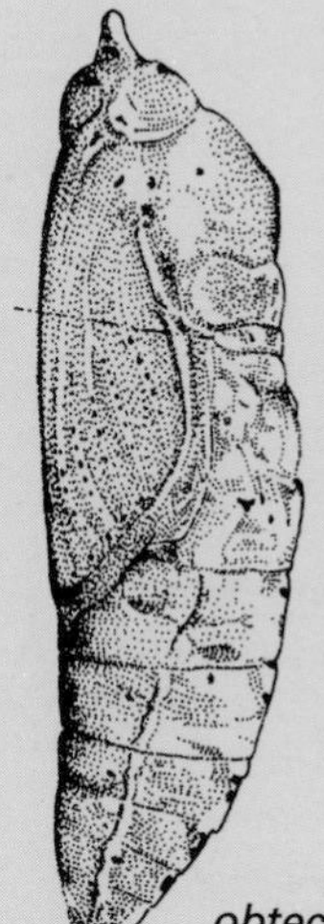
obtect pupa

OBTECT PUPAE

In this type of pupa, the appendages are cemented to the surface of the body by moulting fluid following the final larval moult. The cuticle may be noticeably thickened in the Lepidoptera in particular. The other major order with obtect pupae is the Diptera, but such pupae are also found in two families within the Coleoptera, the Staphylinidae (rove beetles) and Coccinellidae (ladybirds or ladybugs).

03 Classification of Insects

As with almost all groups of living organisms, taxonomists are unable to agree on a universally acceptable classification of the insects. For example, one expert's family may be a different expert's subfamily and there is even greater dissension over which genera particular species may be put into. In the past the placing of insects into groups, whether they be orders, families or genera, has been somewhat inconsistent, often based upon visible characteristics alone. It is not always wise to say that two insects must belong to the same group because they possess a particular structure in common. We now know that similar structures have evolved independently within different groups (convergent evolution) to carry out a similar function. Only if we can show that the structure in the two insects has come from the same common ancestor are we able to say that they belong within the same group. Nowadays insect taxonomists at least agree on the fact that any group must be monophyletic, that is, all of the members of the group must have arisen from a common ancestor. Deciding on the common ancestor at the higher levels of classification is not easy, since in most instances they have been long extinct. With the low levels of fossilization of insects finding these ancestors is not an easy task. As a result, classification relies a great deal upon assumptions and qualified guesswork.

Modern taxonomists are able to call upon a great deal more data than their predecessors, who relied upon external and internal structure and life-cycles

BIOGEOGRAPHY: PROVIDING EVIDENCE FOR THE UNCHANGING FORM OF INSECTS

There is a subtribe of the aphids called the Melaphidina which presently has four separate genera in Asia and a single species in North America. These aphids have a complex life-cycle intimately tied up with plants of the genus *Rhus*, the sumacs. Fossil evidence indicates that their existence in both Asia and America may be ascribed to their being dispersed across the land bridge, which once connected present-day Alaska to Asia. The present distribution of the aphids suggests that their relationship with sumacs evolved before the ancestors of the contemporary hosts became geographically separated into Asian and North American groups. This separation of the sumacs and thus the aphids into two groups occurred about 48 million years ago when climatic changes drove the plants southwards. Eventual separation of Asia and North America by submerging of the land bridge completed the story. Thus, this one group of aphids has been around for at least 48 million years. Perhaps even more astounding is a chironomid midge genus which is found only in Australia and South Africa. Assuming no other answer, this would seem to indicate that the genus existed as long as 150 million years ago at the time when continental drift separated Africa from Australia.

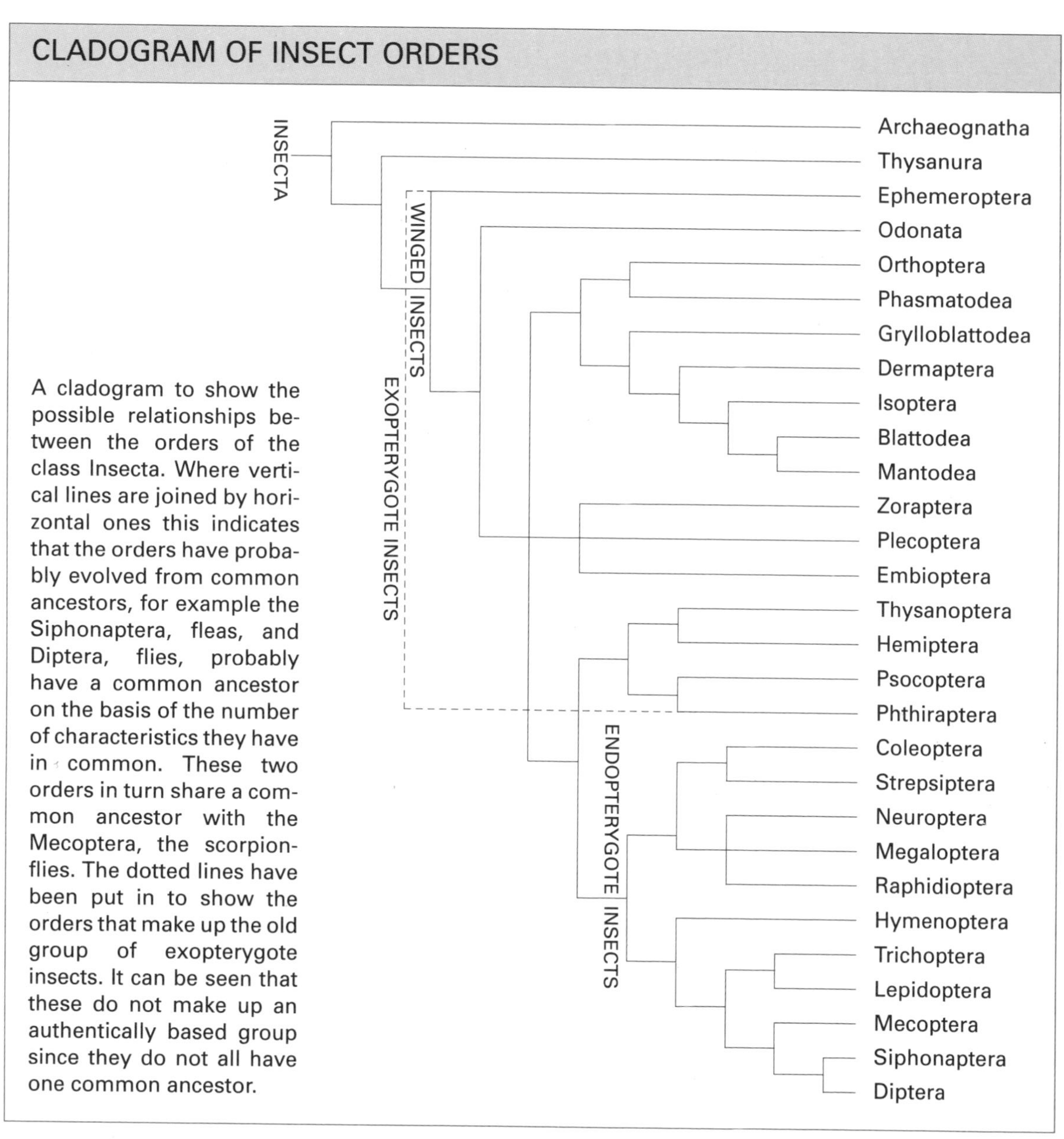

A cladogram to show the possible relationships between the orders of the class Insecta. Where vertical lines are joined by horizontal ones this indicates that the orders have probably evolved from common ancestors, for example the Siphonaptera, fleas, and Diptera, flies, probably have a common ancestor on the basis of the number of characteristics they have in common. These two orders in turn share a common ancestor with the Mecoptera, the scorpionflies. The dotted lines have been put in to show the orders that make up the old group of exopterygote insects. It can be seen that these do not make up an authentically based group since they do not all have one common ancestor.

alone when classifying insects. Today, they can call upon molecular biology and biochemistry to investigate the chemical relationships (especially DNA patterns) between various insects but even this does not always produce satisfactory answers, convergent evolution once again providing problems when interpreting relationships. By using a combination of the old and the new, taxonomists have arrived at a system of classification for the insects whose outline satisfies the majority of entomologists. Before considering this system, however, we will take a quick look at the insect fossil record.

SUMMARY OF THE CHARACTERISTICS OF THE MAJOR ORDERS		
Archaeognatha and **Thysanura**	Bristletails Silverfish	primitive wingless insects, chewing mouthparts and long anal cerci
Ephemeroptera	Mayflies	Chewing mouthparts, two pairs of membranous wings, long anal cerci
Odonata	Dragonflies and Damselflies	chewing mouthparts, two pairs of membranous wings
Orthoptera	Katydids, Crickets and Grasshoppers	chewing mouthparts, leathery forewings covering membranous hindwings, enlarged hind-legs
Phasmatodea	Stick and Leaf Insects	leathery forewings covering membranous hindwings
Dermaptera	Earwigs	chewing mouthparts, short leathery forewings covering membranous hindwings, cerci adapted as pincers
Isoptera	Termites	chewing mouthparts, two pairs membranous wings or wingless in castes, all social
Blattodea	Cockroaches	rather flattened with long antennae, chewing mouthparts, forewings leathery, hindwings membranous
Mantodea	Praying Mantids	chewing mouthparts, long 'neck', raptorial front legs, leathery forewings, membranous hindwings
Hemiptera	Bugs	sucking mouthparts, forewings leathery and or membranous, hindwings membranous
Coleoptera	Beetles	chewing mouthparts, heavily sclerotized forewings and membranous hindwings
Hymenoptera	Sawflies, Bees, Ants and Wasps	chewing and or sucking mouthparts, two pairs of membranous wings which hook together as one
Lepidoptera	Moths and Butterflies	sucking mouthparts, two pairs of membranous wings covered in feathery scales
Diptera	Flies: Crane Flies, Mosquitoes, etc.	sucking mouthparts, first pair wings membranous, second pair as halteres

FOSSIL RECORDS

The earliest fossils which are known to be insects come from deposits in North America which were laid down in the Devonian geological period around 380 million years ago. These were members of the wingless insect order Archaeognatha. By 300 million years ago, in the Carboniferous period, a number of recognizable insect orders existed, most of which are now extinct. Of present-day orders, the Ephemeroptera (mayflies), Blattodea (cockroaches) and Orthoptera (grasshoppers, katydids and crickets) had by then appeared on the scene. This limited number of insect orders may be a reflection of the lack of variety of plant food available to them, since it was not until around 270 million years ago, in the Permian period, that the conifers and allied groups put in an appearance. This resulted in a great increase in the number of insect orders and all of the modern representatives were around by 225 million years ago, with the exception of the Hymenoptera (bees, wasps, ants and sawflies) and the Lepidoptera (butterflies and moths). With the appearance and diversification of the flowering plants from 135 million years ago there was an accompanying increase in the numbers of insects. As might be expected this appearance of the flowering plants was followed by the appearance of the Hymenoptera and Lepidoptera.

The period from 65 million years ago (the time when dinosaurs became extinct) onwards has resulted in some excellent fossils, which have told us a great deal about the antiquity of present day insect genera and species. Even today it is not unusual to find insects trapped in the resin oozing from wounds in the bark of coniferous trees. Where this has happened in the past, the resin has hardened to form amber, containing ancient insects preserved in virtually every detail. These insects in amber have revealed that most modern genera and families are considerably older than was originally thought to be the case. It would appear from the amber that all of the modern insects from the northern temperate, subarctic and arctic zones appeared at least one million years ago. The implications of this are that many insect genera and perhaps even species have been in existence and have remained relatively unchanged for millions of years.

Having delved a little into the ancestry of present-day insects, it is now appropriate to consider each order in more detail. Some of the orders described have a very limited distribution or are rarely seen even by entomologists. Consequently, in the following listings, only those orders most likely to be encountered are in *italic* type.

THE APTERYGOTA (PRIMITIVELY WINGLESS INSECTS)

The only two orders now contained within this grouping, which today are considered as true insects, are:

- THE ARCHAEOGNATHA (THE BRISTLETAILS)
- THE THYSANURA (THE SILVERFISH)

The Apterygota are the most primitive of living insects and their ancestors showed a much greater variation and distribution than their modern descendants. In some older literature these two groups are placed within the same order but this idea is no longer tenable.

ORDER: *ARCHAEOGNATHA* (THE BRISTLETAILS)

Bristletails are mainly creatures of the night and are only likely to be seen by turning over stones, lifting up bark or looking into rock crevices where they spend the daylight hours. When they are found they may be quite abundant. Along the coast of the British Isles, for example, turning over a rock may reveal dozens of *Petrobius maritimus* of all sizes and ages.

The bristletails are a very ancient order of wingless insects. **Petrobius maritimus** is common around the British coast and may often be found in large concentrations. When threatened it can run rapidly and can even jump some distance.

Bristletails feed on various types of dead and decaying plant material, such as algae, lichens and mosses. When disturbed, they run very rapidly and have the ability to jump quite long distances by arching their body and then suddenly straightening up.

Apart from a lack of wings, they are mainly characterized by their long, multi-segmented, threadlike antennae, their very long maxillary palps, large compound eyes, the humped thorax and the paired cerci, which are noticeably shorter than the median caudal appendage or 'tail'. Microscopic examination shows that each mandible possesses only a single condyle (articulation) onto the head and that along the abdomen on segments II to IX there are styles representing primitive abdominal limbs. The whole body is covered with pigmented scales.

Males deposit a spermatophore onto the substrate and this is then picked up by the female. Development occurs through a series of moults with each instar a small replica of the adult.

ORDER: *THYSANURA* (THE SILVERFISH)

The term 'silverfish' should strictly speaking be confined to the family Lepismatidae, which contains the familiar silverfish found in human dwellings, though there is now a tendency to use the name for the whole order. They are nocturnal living under bark, in litter, in termite and ant nests, mammal burrows, caves and also, of course, with man. They feed mainly on plant material and can be a pest in our homes when they attack bookbindings, wallpaper, etc.

Silverfish superficially resemble bristletails but are more flattened and they cannot jump. Their antennae are multisegmented and threadlike but the compound eyes are reduced or absent. The caudal appendage and the paired cerci are approximately equal in length. A major advance on the Archaeognatha is that each mandible has two articulations with the head, permitting a transverse biting action of one mandible against the other rather than the less efficient rolling action of the bristletail mandibles. Abdominal styles may be present on segments II to IX though these are more often on segments VII to IX only.

Males again produce a spermatophore to be picked up by the female and development is by a series of instars similar to the adult. Some species are parthenogenetic, that is, they reproduce in the complete absence of any males. Unusually for insects they can live quite a long time, up to four years in some species.

THE PTERYGOTA (WINGED OR SECONDARILY WINGLESS INSECTS)

The Pterygota contains the rest of the insect orders in existence today. Two main subdivisions of the group, based upon the way the wings may be folded, seem to be generally accepted by today's entomologists:

- ♦ PALAEOPTERA
- ♦ NEOPTERA

The Palaeoptera includes the mayflies and the dragonflies and damselflies, which are unable to fold their wings against the body when at rest – they are the most ancient of the flying insects. The Neoptera contains all the remaining insect orders.

Palaeoptera

ORDER: *EPHEMEROPTERA* (THE MAYFLIES)

Although they have a worldwide distribution, with the exception of Antarctica and some oceanic islands, the Ephemeroptera reach their greatest variety and numbers in the world's temperate zones. The adults are free-flying whereas the nymphs are almost without exception aquatic (there is a single species with terrestrial nymphs).

The adults are easily recognizable with their large, transparent forewings held vertically over the body and their two or three fine, caudal appendages extending from the rear of the abdomen. A closer examination reveals that the adults have compound eyes, often large especially in males, three ocelli and small, slender antennae. The mouthparts of the adult are very much reduced for it does not feed and spends its very short life solely in looking for a mate and in reproduction. Males have noticeably elongated forelegs which they use for grabbing hold of the females during nuptial flights. The forewings are very large but the hindwings are very much reduced in size. The larvae somewhat resemble wingless adults. Since they are the only feeding stage, however, they do possess jaws with which they feed upon fine vegetable matter such as encrusting algae, though a few species have predatory larvae.

The larvae have the three caudal filaments seen in the adult but they also possess pairs of feather-

Although at first sight this appears to be an adult this is in fact a sub-imago or 'dun' of the mayfly, **Ephemera danica**, and it has to undergo a further moult before it becomes fully adult. Characteristic of the order Ephemeroptera are the wings held above the body, since they have no folding mechanism, and the three caudal appendages trailing from the abdominal tip.

like gills on some of the abdominal segments. In later instars the developing wings clearly protrude from the thorax. The Ephemeroptera are unique in having a subadult subimago stage. This emerges from the final aquatic stage, flies to waterside vegetation and then within a variable period of time, depending on species, moults to become a sexually mature adult. Like the adult, the subimago does not feed. The male mates with the female in a mating flight and introduces sperm directly into her body. In exceptional cases, mating takes place in the subimago stage and there is no adult stage. In a number of species the females can reproduce by parthenogenesis and in at least five species this is obligatory since males are not known to exist.

ORDER: *ODONATA* (DRAGONFLIES AND DAMSELFLIES)

The Odonata comprise two suborders:

- ANISOPTERA
- ZYGOPTERA

The dragonflies (suborder Anisoptera) and damselflies (suborder Zygoptera) are perhaps more familiar to us than the mayflies on account of their larger size. The smallest of the Odonata are around 20mm in length whereas the largest attain a body length of 150mm with a wingspan up to 190mm (the long extinct *Meganeura* had a wing span four times this figure). The Odonata are fast-flying (up to around 95km/hr), predators which catch their prey on the wing. They are found in most parts of the world as long as there is freshwater in which they can breed. The adults are found around water during the breeding season but otherwise they may hunt at considerable distances from water, even penetrating some distance into deserts to feed. Some species are migratory.

In the field, a number of characters distinguish the dragonflies from the damselflies. In general, the dragonflies are larger and more robustly built than the damselflies, which also tend to have a much weaker flight. Closer inspection reveals that the hindwings of dragonflies are much wider at the base than the forewings and both wings are held out laterally at right angles to the body when at rest. In damselflies, both pairs of wings are almost identical in shape and more often than not they are held against one another above the abdomen when at rest. A small Asian suborder, the Anisozygoptera, with just two described species combines characteristics of both main suborders.

Odonates have a very mobile head, a feature which is very obvious when watching one at rest in the wild. The head bears a pair of small antennae, two very large compound eyes, three ocelli and well-developed chewing mouthparts. The thorax is large to contain the powerful flight muscles and the abdomen is long and slender, though broader in the libellulid dragonflies. The larvae of Odonata are aquatic predators with an extensible labium, the 'mask', which can shoot out and grab passing prey. Larvae of the two main suborders show similar differences in build as the adults, though the three flattened caudal gills on the damselfly nymph easily distinguishes them from dragonfly nymphs. The Odonata are unique amongst the insects in that the male copulatory apparatus is situated on abdominal segments II and III rather than on segment IX. Segment IX still has the opening from the testes so that, prior to mating, the male has to bend the abdomen round to transfer semen from the testes to the penis vesicle, by bending the end of the abdomen round to bring the two into contact. This accounts for the peculiar mating stance, the 'wheel position' adopted by the Odonata, where the female has to bend the end of the abdomen round to bring the genital opening into contact with the front end of the male's abdomen.

Neoptera

In all of the remaining insect orders the wings are capable of being folded flat over the body. As noted, the Neoptera may be subdivided into two groups, on the basis of the developmental progression from egg to adult. In the exopterygote insects there is a series of nymphal stages, as in the Palaeoptera, each resembling a miniature, wingless adult to a greater or lesser degree. The wings develop gradually on the outside of the nymphal thorax. In the endopterygote insects, there is a full metamorphosis. From the egg there hatches a larva which seldom bears any resemblance to the adult. After several moults, this forms a pupa in which the larval organs break down and are reprocessed to give rise to the adult organism. The adult insect, the imago, then emerges from the pupa.

In older literature the exopterygote insects are actually classified within the Exopterygota (or Hemimetabola), but this grouping is now considered untenable, since it does not represent a proper taxonomic unit, some members within the grouping approaching the endopterygote life-cycle with a pupa-like stage in it. The Endopterygota (or Holometabola) are, however, acceptable since they all possess in common a resting pupal stage in the life-cycle.

Exopterygote Insects

ORDER: *ORTHOPTERA* (KATYDIDS, CRICKETS AND GRASSHOPPERS)

This is a very familiar insect order. Some confusion has, however, arisen within the common names, in that in certain countries katydids are more often referred to as 'bush-crickets' and worldwide some, but by no means all, of the larger swarming grasshoppers, are called 'locusts'.

The Orthoptera are subdivided into two suborders:

- ♦ CAELIFERA
- ♦ ENSIFERA

The Caelifera are the grasshoppers and the Ensifera contains all of the others. The compound eyes are large, ocelli may be present or absent and the jaws are well developed for cutting and chewing. The pronotum is well defined and covers not only the top but often extends well down the sides of the prothoracic segment. The first pair of wings, when present, are leathery and act as covers for the second pair of membranous flying wings, which fold up fan-like beneath the forewings when at rest. The most noticeable orthopteran characteristic, however, is their enlarged hindlegs which are used for jumping. Many male orthopterans have sound producing (stridulatory) organs, each species producing a unique 'song' to attract a mate.

The Caelifera are mainly diurnal plant feeders, often brightly coloured, jumping and flying freely when disturbed, though mountain and rainforest dwelling species often have reduced wings and cannot fly.

The Ensifera have both diurnal and nocturnal members; though many feed on plants a number are carnivorous. Of the plant feeders, a number burrow in the ground to feed on roots. Some are even cave-dwellers. Many katydids have fully developed wings, though these are often reduced in members of the other ensiferan groups.

CAELIFERA AND ENSIFERA

There are two major points which distinguish Caelifera from Ensifera in the field. The antennae of grasshoppers are very short, only just exceeding the length of the head in most instances, whereas those of the katydids and their allies are much longer, sometimes more than twice the total body length. Thus in some older literature, the Caelifera are called 'short-horned grasshoppers' whereas the Ensifera are the 'long-horned grasshoppers'. Female Ensifera have a long ovipositor extending from the tip of the abdomen whereas that of the Caelifera is very short and is not normally visible. Orthopteran nymphs look like miniature adults, with a relatively larger head in the earlier instars, though they may differ markedly in coloration from the adults.

At one time the Phasmatodea (*see* below) were included within the Orthoptera but they are now considered to be the latter's sister group and sufficiently different to merit their own order.

ORDER: *PHASMATODEA* (STICK INSECTS AND LEAF INSECTS)

The stick insects, or walking-sticks, are familiar since they are often kept in schools or by children as pets. As their name implies, most stick insects have a small head, long thin thorax and abdomen and proportionately long legs. The largest species can grow up to 300mm in length. A number of species are, however, much shorter and quite squat in appearance. Leaf insects, on the other hand, have shorter, dorsoventrally flattened bodies with a marked resemblance to the leaves amongst which they live. All Phasmatodea are vegetarians and are mainly active at night, their resemblance to sticks or leaves protecting them during daylight hours.

The head bears chewing mouthparts, small compound eyes, ocelli in flying species (more often in males) and relatively short antennae. The mesothorax and metathorax tend to be more elongated in those species which fly. In the latter, the forewings are leathery and protect the membranous hindwings, whose forward margins are toughened to protect the wings when folded. The nymphs are like miniature adults.

The stick-like form of the average member of the order Phasmatodea is evident in this mating pair of an unnamed species from Trinidad. The male stick insect, as in this instance, is considerably smaller than the female and often differently coloured or marked. The female in particular of this species resembles the stems of the vine, which can be seen adhering to the trunk of the tree on which they live.

ORDER: GRYLLOBLATTODEA (ROCK CRAWLERS)

A tiny order with just twenty species in a single family, the rock crawlers are inhabitants of cold, mountainous areas of western North America, Japan and mainland Asia, where they live among stones or in caves, including ice caves. In shape they bear a resemblance to the familiar earwig without the forceps on the end of the abdomen. The compound eyes are small or absent and there are no ocelli. Chewing mouthparts are present with which they feed on dead arthropods and other organic matter, which they scavenge from the surface of ice and snow, especially during the spring melt. The legs are well developed for running but wings are absent. There is a pair of long cerci on the end of the abdomen and in females a short, sword-like ovipositor rather like that of female katydids in appearance.

ORDER: *DERMAPTERA* (EARWIGS)

The most obvious feature of the earwigs is the way in which the cerci at the end of the abdomen are modified to form a pair of forceps or pincers, which are usually at their most robust in males. Other-

The most obvious characteristic of the order Dermaptera, the earwigs, is the way in which the abdominal cerci are modified to form a pair of pincers. This is a desert-dwelling species, **Labidura riparia**, and as well as the pincers the distinctive short wing cases are plainly visible.

wise, they have short antennae, compound eyes varying between large and absent, no ocelli and chewing mouthparts. The forewings are small and leathery and cover the folded hindwings, which may be used for flying, though earwigs seldom seem to do so, running being the preferred method of locomotion for most.

The nymphs are small replicas of the adults and all stages of most earwigs are scavengers, feeding upon whatever dead and decaying animal and vegetable matter is available. Two families of rather unusual earwigs are associated with mammals and may be semi parasitic or fully parasitic on them.

Despite the fact that their name is supposed to be derived from their habit of crawling into people's ears, there is no proof that this is anything other than a rare accidental occurrence.

ORDER: *ISOPTERA* (TERMITES)

All living termite species are social, living in colonies which number from hundreds for some species up to millions for others. Their sociality, like that of the bees, wasps and ants is based upon a caste system, the different castes often having marked structural adaptations. The basic termite has chewing mouthparts, fairly long antennae and compound eyes, which may be reduced or absent. There may be bizarre modifications of the head in some soldier termites relating to their role of colony protectors. Both pairs of wings, when present, are membranous and are similar in size and shape (except in *Mastotermes*). The wings have a weak fracture point at the base and are shed following dispersal flights, when they are no longer needed. Cerci are present but

Unless the nest is destroyed to uncover the king and queen termites, the individuals of any species of the order Isoptera most likely to be encountered are foraging workers and their attendant soldiers. Here, large headed soldiers of a **Macrotermes** species are guarding the smaller workers, which are harvesting dead leaves in a Malaysian forest. Although sometimes called 'white ants' on account of their behavioural similarities, the obvious difference between the ants and the termites is the latter's lack of a narrow waist between the thorax and the abdomen.

small and with one exception (*Mastotermes*) there are no external genitalia. They feed mainly on plant material and to aid in its breakdown their guts contain symbiotic cellulose-digesting bacteria.

ORDER: *BLATTODEA* (COCKROACHES)

This group would no doubt be unfamiliar if it was not for the notoriety of a few members of the order regarded as pests. They are rather flattened insects, with long, slim antennae, compound eyes varying from the moderately large down to the absent and chewing mouthparts. The pronotum is well developed, the forewings are leathery, acting as covers for the membranous hindwings and the legs are long and used for running. The abdominal cerci are well developed.

Typical of the exopterygote insects, the nymphs resemble the adults, with gradual development of the wings externally visible. While many species are nocturnal, others are diurnal and they cover a range of habitats. Some live on the ground, others on plants, including trees and a few are cave-dwellers. They are mainly scavengers, feeding upon whatever suitable organic matter is available to them. A few species have single-celled organisms in their guts to aid digestion.

The North American woodroaches of the genus *Cryptocercus* are of particular interest in that the adults exhibit parental care, taking up to six years to rear a single brood on their nutrient-poor diet of decaying wood.

From the eucalypt forest of Queensland in Australia comes this **Ellipsidion** sp. cockroach, a member of the order Blattodea. Characteristic of this order is the broad, flattened body form, the long, many-segmented filiform antennae, the broad pronotum covering the thorax and the front end of the abdomen and the way in which the forewings, acting as covers for the membranous hindwings, overlap one another along their length. This latter feature is not found in beetles, to which cockroaches bear a superficial resemblance.

ORDER: *MANTODEA* (PRAYING MANTIDS)

This order has gained media attention on account of the fact that female mantids are supposed to eat the male following mating. Although this may often happen under captive conditions, it seems to occur only rarely in the wild. Mantids are medium-to-large insects, with females of some species attaining 150mm in length, though males are usually smaller than females. They have small, highly mobile heads with slim antennae, relatively large compound eyes and chewing mouthparts. The prothorax is long and narrow, giving the appearance that the head is on the end of a long neck. In winged individuals the forewings are leathery and cover the membranous hindwings. The forelegs are raptorial, that is, they are modified for grasping prey, while the other two pairs of legs are fairly long and used for walking and running. Cerci are visible on the end of the abdomen.

Nymphs resemble adults, but lack wings, and all stages are 'sit-and-wait' predators, using their excellent eyesight to pick out, and their raptorial front legs to grasp, any suitable arthropod prey which comes within range.

ORDER: ZORAPTERA (ZORAPTERANS)

This is a tiny order with a single genus, *Zorotypus*, containing some thirty species only. They are mainly from the warmer parts of the world and their habit of living in leaf-litter, rotting wood or around termite nests means that they are highly unlikely to be encountered by anyone except the most determined enthusiast. They are gregarious insects with chewing mouthparts and with compound eyes and ocelli present in winged individuals, absent in those without wings. Like the termites, the wings have a basal fracture line and can be shed.

ORDER: EMBIOPTERA (WEB-SPINNERS)

Like the previous group, the web-spinners are gregarious insects but, again, they are unlikely to be seen even by the most enthusiastic naturalist. Their common name is derived from their habit of building silken tunnels in which they live, making good their escape along these by running forwards or backwards with equal agility.

They are rather elongate insects with kidney-shaped compound eyes, chewing mouthparts and wings present in the males only. The males are most likely to be encountered since they are attracted to lights at night. They live in leaf litter, beneath stones or bark or cracks in the ground. As one lot of food becomes exhausted so they extend their silken tubes out to cover new sources of moss, dead leaves, bark and other plant litter on which they normally feed.

Caught in the act of 'prayer', from which the order Mantodea gets its general common name of praying mantids, is an adult male of **Polyspilota aeruginosa** from East Africa. Note how the mobile head, with its chewing mouthparts and large compound eyes, is situated at the end of a long 'neck' formed by the prothorax. The raptorial front legs, with their long spines, are held out ready to grasp any potential prey item. This is quite a large insect and the front leg spines are capable of piercing human skin if the mantid is picked up carelessly.

ORDER: *PLECOPTERA* (STONEFLIES)

This is a group of insects whose adults, if they are to be found at all, will not be far from a source of clean, running water or a clear lake. They are most likely to be familiar to trout fishermen, since stonefly nymphs and adults are fed upon by trout and form the basis of a number of 'tied flies'. Adults have long, slim antennae, compound eyes, ocelli, chewing mouthparts (when present) and long wings, which are held flat against and extend beyond the end of the body in fully winged individuals. Both pairs of wings are membranous, though the hindpair are usually the broader of the two. There is usually a pair of long cerci on the end of the abdomen. Though just a few are terrestrial, the majority of stonefly nymphs are aquatic, obtaining oxygen through the whole body surface or with the aid of gills situated on various parts of the body.

The nymphs may be detritus feeders, plant feeders, predators or omnivores, that is, a combination of these. Adults have weak mouthparts and are known to feed on algae, lichen, dead wood or even pollen. The final nymphal stage crawls out of the water before the adult emerges.

ORDER: PHTHIRAPTERA (LICE)

All of the members of this insect order are parasites upon birds and mammals. It includes the familiar human headcrab and body lice as well as the feather lice, all too familiar to poultry farmers. Living as they do amongst fur and feathers, these parasites have little need to see anything and their eyes are consequently reduced in size or absent altogether and they lack ocelli. Long antennae would hinder these insects and these are much reduced in length.

There are four suborders of Phthiraptera:

- ANOPLURA
- ISCHNOCERA
- AMBLYCERA
- RHYNCHOPHTHERINA

The mouthparts are either adapted for piercing and sucking (Anoplura) or for chewing (Ischnocera and Amblycera). The Rhynchophtherina contains but two species in a single genus. They are usually called elephant lice, since one species lives on elephants, though the second species actually lives on the warthog. All of the lice are generally flattened, wingless insects, with short legs. The chewing lice feed upon feathers and dead skin, with a few adapted to feeding on blood, which they release by chewing holes in the skin. All of the Anoplura, the biting lice, feed by piercing the skin and sucking the host's blood.

ORDER: PSOCOPTERA (BARKLICE AND BOOKLICE)

Rarely exceeding 5mm in length, these tiny insects are likely to be overlooked or misidentified. They have round, mobile heads with large compound eyes and ocelli in winged forms. The antennae are long and slim and the mouthparts are used for chewing. Some species are wingless while others possess a pair of membranous wings. The forewings are larger than the hindwings and they are hooked together during flight. When at rest, the wings are held tent-like over the body. There are no cerci on the abdomen.

Both adults and nymphs feed on fungi, algae and lichens while others are associated with man,

Members of the order Psocoptera that are most likely to be met with are tiny creatures 2–3mm in length, brownish yellow in colour which secrete themselves in nooks and crannies in human habitations. Supporting itself by scraping the surface from leaves, however, is this attractive member of the order form Peru. When present, as here, the wings are held in a typical tent-like attitude over the abdomen.

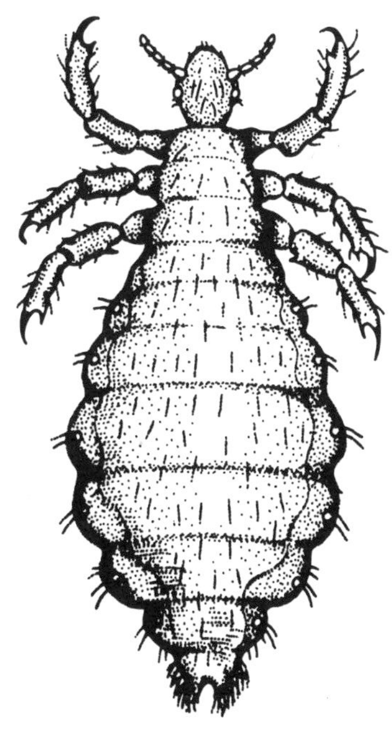

A human head louse, **Pediculus humanus** showing the hooks on the ends of the legs with which it hangs onto the hairs.

feeding on paper, book-bindings and cereal products, reaching pest proportions on occasions. A number of species are gregarious and live on trees and bushes on a silken sheet. Parthenogenesis is common amongst the Psocoptera. In some species it is the norm; in others normal reproduction with males may also occur.

ORDER: *THYSANOPTERA* (THRIPS)

Like the two preceding orders, the thrips are tiny insects seldom encountered except when they become a nuisance. The largest species may attain a body length of 15mm but the majority are much smaller than this. They are slim bodied with short antennae, small compound eyes and ocelli are present in winged forms. The mouthparts are adapted not for chewing but for piercing and sucking. The wings, when present, are most unusual in that each one consists of a membranous, strap-like central portion with a setal fringe down either side, which forms the main aerodynamic surface. The wings are roughly identical in size and shape, and on each side interlock by means of a row of hooks at the base of the hind wing, which engage with a fold on the back edge of the forewing.

The initial stage of development involves nymphs resembling adults, but without wings. The final stages, however, include two resting stages, the prepupa and the pupa, during which there is some structural breakdown and then reconstruction, in a manner similar to that of the endopterygote insects with their larval and pupal stages. The majority of thrips species feed by piercing plant cells and sucking out the cell sap; consequently, they include some pests. A few thrips are predators.

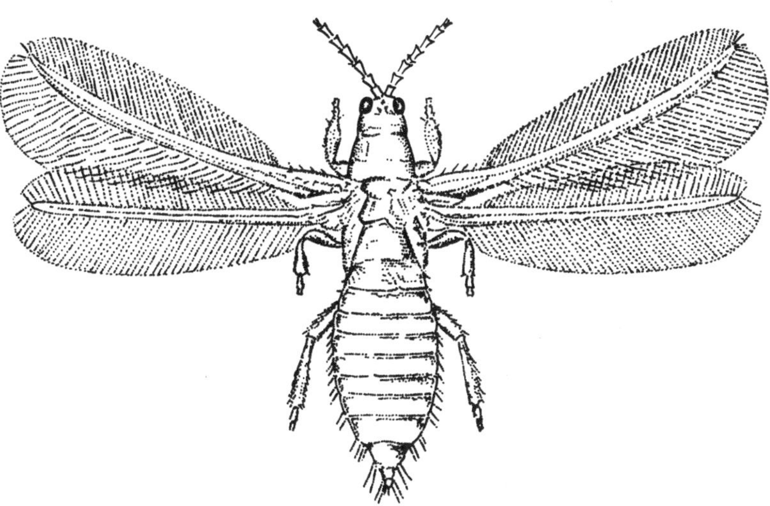

Because of their small size thrips need to be observed closely under a microscope, when their characteristic feathery wings become apparent.

ORDER: *HEMIPTERA* (BUGS)

Although there is a tendency nowadays to refer to any creepy-crawly as a 'bug', the true bugs are all members of the order Hemiptera. Traditionally they are divided into two suborders:

- HETEROPTERA
- HOMOPTERA

The Heteroptera are simply referred to as 'bugs' and the Homoptera are more usually called 'plant bugs'.

In common, the two suborders have the following characteristics. They vary greatly in general shape from the flattened disc of the bed bug to the stick insect-like water measurer. Compound eyes are present and these are often large; ocelli may or may not be present. The antennae are simple but vary greatly in length and the number of segments present. Mouthparts consist of a rostrum or proboscis which is used for piercing and sucking food items. Walking legs are present but the forelegs may be raptorial in some forms, for example, some assassin-bugs and the water scorpion. Two pairs of wings are often present, though in some species these may be reduced or absent. Some species of bugs have fully winged (macropterous), reduced winged (brachypterous) and unwinged (micropterous) forms in the adult stage. In male scale insects, only the forewings are present while their females are completely wingless.

The Heteroptera have the ability to swing their proboscis both down and forward from its resting position beneath the thorax and abdomen. The base of the forewings is hardened, though the rest is membranous, while the hindwings are completely membranous. The wings are held flat over the body at rest with the membranous areas overlapping. Abdominal scent glands, often called 'stink glands' are always present. Many Heteroptera are plant feeders, either piercing plant stems and tapping the sap flow or feeding on seeds, while many more are predators, feeding on other arthropods by piercing them and sucking out their body contents.

Development is through a series of nymphs which may resemble the adults, but without wings, to a greater or lesser extent depending on the species. Some, for instance, have brightly coloured nymphs but drab adults while in others the nymphs may have drab, cryptic coloration while the adults possess bright, warning coloration. A number of nymphs mimic ants while the adults are normal-looking bugs.

The Homoptera lack the ability to swing their proboscis forwards and as a consequence they are restricted to feeding on plants. Both pairs of wings, when present, are membranous and are held roof-like above the body. The life-cycles of the Homoptera are less straightforward than those of the Heteroptera. Many aphids, for example, reproduce parthenogenetically for most of the time but have a sexual generation in one particular season as well. In whiteflies and scale insects, the nymphs resemble the larvae of the endopterygote insects and there is a resting stage akin to the pupal stage of the latter group. This characteristic relates the bugs to the thrips and makes these two groups possible intermediates between the exopterygote and the endopterygote insects which follow.

The Endopterygote Insects

ORDER: *COLEOPTERA* (BEETLES)

In terms of variety, the Coleoptera are the most successful animals on Earth, with over 300,000 species already described. They vary in size from the giant hercules beetles, the world's heaviest insects, measuring up to 155mm in length, down to tiny species only 0.5mm long. Despite this range in size, they have a certain uniformity which enables them to be instantly recognizable as beetles. Adult beetles are often heavily sclerotized, some to the extent that they are truly armoured. Most adult beetles have well-developed compound eyes, though a few species living in the permanent darkness of caves and similar habitats lack eyes altogether; ocelli are usually absent. Beetle antennae are very variable, both in the number of joints that make them up, their arrangement and their length. At one

Psellopius rufofasciatus from Mexico is an assassin bug belonging to the suborder Heteroptera, the true bugs, within the order Hemiptera. Like all Hemiptera, it has sucking mouthparts, in this instance to plunge into and suck out the body contents of its arthropod prey. Distinguishing it from the Homoptera are the forewings, which lie flat over the abdomen and are partially sclerotized, with only the hind end of each being membranous and overlapping the adjacent wing.

This odd-looking creature is a treehopper of the family Membracidae from Trinidad. It is a member of the suborder Homoptera, within the order Hemiptera, characterized by their sucking mouthparts used to extract sap from plants. Homoptera differ from the Heteroptera in having both pairs of wings membranous and held tent-like over the body, not overlapping. This particular species, **Cyphonia clavata**, may well be mimicking an ant with the peculiar extensions on the rear end of its pronotum.

extreme they may be one or two jointed, at the other up to twenty-seven or more. They may be more than twice the body length in some longhorn beetles, though in most beetles they are shorter than the body. Only within the family Meloidae, the oil-beetles, do we find an exception to the beetles' possession of chewing mouthparts. These meloids exceptionally have their mouthparts adapted for sucking up nectar from flowers. There may be a great enlargement of the mandibles in some male stag beetles of the family Lucanidae, to the extent that they may exceed the body in length.

The forewings of beetles are sclerotized to form rigid covers, the elytra, to protect the membranous hindwings. Although the elytra may be used to help in lifting off, they are not used in flight but are held out sideways clear of the hindwings. The elytra are fused together and the hindwings are absent so that flight is not possible in many weevils and some ground (Carabidae) and darkling beetles (Tenebrionidae). In most beetles, the elytra cover the thorax and abdomen but in the family Staphylinidae they cover the thorax only. In these latter beetles the hindwings have a special folding pattern to fit beneath the small elytra. Beyond being used for walking and running, the legs may be adapted for digging, swimming or, as in the flea beetles, jumping.

Beetle larvae also show a great deal of variation but in common they have a sclerotized head capsule bearing a pair of mandibles and up to six pairs of stemmata. Usually three pairs of thoracic legs are present though weevil larvae lack these. The pupa is covered by a thin, soft cuticle and is often found beneath the ground or inside the beetle's food plant. A number of pupae are contained within cocoons, though these are rarely of silk.

Adult beetles and their larvae are found in almost every habitat, though like most insects, few of them are marine. They feed upon almost every kind of organic material that is available. Many are straightforward plant feeders or predators, others feed on fungi, dung and carrion while just a few are parasitic. With such a wide range of feeding habits it is inevitable that they include many pests of man's crops and stored food products.

With the front of the head extended into a rostrum or snout, which in some species can approach the body in length, the weevils are perhaps less beetle-like than the average member of the Coleoptera. This is **Cholus cinctus** feeding on a Heliconia flower in rainforest in Costa Rica.

ORDER: *STREPSIPTERA* (STYLOPS)

This is a small insect order, with around 400 species, all of them parasites of other insects, notably Hymenoptera and Hemiptera. As a result of the similarity of their larvae to those of certain beetles, they have in the past been included within the Coleoptera but modern thinking places them within their own order. The males and females are so different in appearance that it is difficult to accept that they are in fact one and the same creature. The males are recognizable as insects, even though they have certain peculiar features. The head is much broader than long and the compound eyes are large, though they have few facets. The antennae are short and may be branched and the chewing mouthparts are much reduced. The forewings are reduced to stubs while the hindwings are fan-shaped. The metanotum is developed backwards to cover the anterior end of the abdomen. The female is grub-

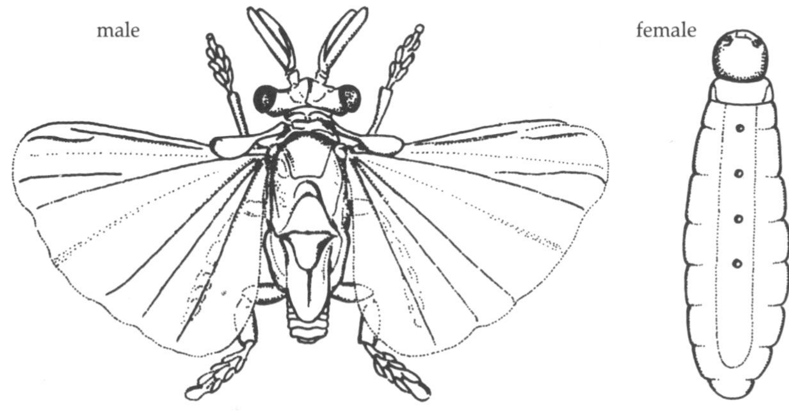

Stylops.

like in appearance and remains within the host, inside the puparium formed from the last larval cuticle, and is sought out by the short-lived male in order for mating to take place.

Eggs hatch within the female to release the first instar triungulin larvae. These have no mouthparts or antennae but do have well-developed eyes and three pairs of thoracic legs. They leave their host whilst it is perched on a flower and await the arrival of a new host. They then cling onto this to get a ride to its nest where they penetrate one of the host's eggs or larvae. The triungulin larva then moults to become a legless, maggot-like second instar, which feeds by absorbing nutrients directly through its cuticle from the blood circulating within the host's body cavity. Pupation takes place within the host, the pupa protruding from the host's body wall, giving an exit route for the males when they emerge.

Although making a good attempt to mimic a polistine social wasp this **Climaciella** sp. mantispid from Brazil is in fact a member of the order Neuroptera. It differs from the wasp and other Hymenoptera in that the wings are almost equal in size whereas in a wasp the hindwings are smaller, the wings have cross-veins to give a net-like appearance not seen in wasp wings and the 'neck' is the extended prothorax, a character not met in the Hymenoptera. The raptorial front limbs in the mantispids are used for grasping prey.

ORDER: *NEUROPTERA* (LACEWINGS, MANTISPIDS AND ANT-LIONS)

Although a fairly large order with around 4,500 species worldwide, the Neuroptera are generally unfamiliar, although in northwest Europe the green lacewing is a common insect, since it often enters houses to hibernate during the winter months. The predatory habits of the ant-lion larva are also well documented. Adult Neuroptera have large eyes and mouthparts developed for chewing. The wings are similar in shape and venation, with the exception of some ant-lions, whose hindwings are modified as streamers, the forewings alone being used for flight. The abdomen bears no cerci.

The family Mantispidae, as their name implies, resemble praying mantids to some degree. The prothorax is elongated to form a long 'neck', the head is mobile and they have raptorial front legs for snatching prey. Although many adult Neuroptera are predatory, a number feed on nectar and pollen. Larval Neuroptera are predatory and have the chewing mouthparts elongated so that they can pierce the body of their prey and suck out the body contents. Larvae and pupae are mainly terrestrial.

ORDER: MEGALOPTERA (ALDERFLIES, DOBSONFLIES AND FISHFLIES)

Included by some authorities within the Neuroptera, and closely related to them, this small order of only around 300 species is most likely to be familiar to fishermen, since the adults are found around water and the larval stages are aquatic. Adults resemble the Neuroptera in appearance but never have the elongated 'neck' or adapted front legs or hindwings. Although the adults have well-developed jaws, they are not used in feeding; males have small cerci on the abdomen.

ORDER: RAPHIDIOPTERA (SNAKEFLIES)

Named for their slim, elongated, snake-like 'neck', the Raphidioptera are again included in the Neuroptera by some authorities. This is a very small order with only about 80 species within a single family. Both adults and larvae are terrestrial predators and the adults strike at their prey, using the long 'neck' in a snake-like way. Both pairs of wings are membranous, small cerci are present and the female has a long egg-laying tube on the end of her abdomen.

A female **Agulla** species female Snake-fly showing the characteristic long 'neck' of this order. The needle-like extension from the end of her abdomen is her ovipositor.

ORDER: *HYMENOPTERA* (SAWFLIES, BEES, ANTS AND WASPS)

This is a large order with many members of economic importance, either as friends or foes, to us humans. More than 120,000 species have so far been described, though twice this number probably exist, and they are found in all parts of the world with the exception of the Antarctic region. The Hymenoptera includes some of the world's smallest insects, parasitic wasps only 0.15mm body length. At the other end of the scale some ichneumon wasps may attain a body length of 120mm.

Adult Hymenoptera have large, compound eyes and ocelli may be present. Mouthparts are mainly for chewing though in the bees they have become modified for sucking. Adults generally feed on nectar or honeydew or, in the case of some hunting wasps, on a portion of the body fluid of the prey that they catch to feed their offspring. Both pairs of wings are membranous, the forewings larger than the hindwings. Along the leading edge of the hindwing is a row of hooks (hamuli), which couple it to the trailing edge of the adjacent forewing during flight.

The Hymenoptera is subdivided into two distinct suborders:

- SYMPHYTA
- APOCRITA

The more primitive is the Symphyta (sawflies, horntails and wood wasps) and the more advanced the Apocrita (parasitic wasps, wasps, ants and bees). The Symphyta have a broad attachment between the thorax and abdomen, there being no narrow 'waist', which is a characteristic of the Apocrita. Female Symphyta usually have a saw-like ovipositor used for inserting eggs into plant tissue. The characteristic 'waist' of the Apocrita is formed by incorporation of the first abdominal segment into the rear of the thorax. Larval Symphyta, with the exception of wood-borers, have segmented thoracic legs but larval Apocrita are legless.

The Apocrita are themselves divided into two distinct divisions:

- PARASITICA
- ACULEATA

The parasitic wasps (Parasitica) have an ovipositor which they use to deposit their eggs in or on their host animal or plant. Larvae develop in or on the host. Females of the wasps, ants and bees (Aculeata) have the ovipositor modified to form a sting used as a defensive organ, for killing or subduing prey, or both of these, though the stings have been secondarily lost in a number of ant species.

ORDER: *TRICHOPTERA* (CADDIS-FLIES)

On account of their purely aquatic larvae the caddis-flies are always associated with rivers, streams, lakes and ponds. The adults are somewhat moth-like in appearance and indeed the Trichoptera are close relatives of the Lepidoptera. Adult caddis-flies have large compound eyes, two or three ocelli, long, thread-like antennae and reduced mouthparts, since they do not feed, though some may drink nectar or water. Both pairs of wings are membranous and like those of the Lepidoptera, they may be scaled. The larvae have fully developed chewing mouthparts and three pairs of thoracic legs.

Many people associate caddis larvae with the case in which they live. This may be made of sand, pebbles, twigs or other detritus depending upon where they live. The larva protrudes its head and thorax from the case when moving or feeding, withdrawing instantly into its protecting walls when threatened. Some larvae are, however, free-living while others live within a silken net, which acts as a snare in running water. The larvae may be predators, they may filter out or shred organic matter or they may be grazers on plants. The pupa is also aquatic and forms within a pupal case, either part of the original larval case or a specially constructed shelter. The latter is made by free-living larvae prior to pupation, which takes place within a cocoon. When development is complete the pupa uses its mandibles to bite its way out of the case before swimming or walking to the surface, where the adult emerges.

ORDER: *LEPIDOPTERA* (MOTHS AND BUTTERFLIES)

The moths and butterflies require little in the way of introduction since they are, along with hive bees, the most familiar of insects to many of us. With around 150,000 species so far described, the Lepidoptera is a large order whose members are found worldwide. The separation of moths from butterflies is something of an artificial one, since the skipper butterflies form a link between the typical moth and the typical butterfly. Also, there are

Taking up salts from the rainforest floor in Peru is this nemeobiid butterfly, **Rhetus periander laonome**. As a member of the order Lepidoptera, its most obvious distinguishing features are the rows of scales covering the wings and the proboscis curving away below the head. This wing posture is more typical of butterflies while the majority of moths at rest fold the wings over the body in a tent-like fashion.

BUTTERFLY OR MOTH?

In general, butterflies are regarded as day-flying Lepidoptera, with filamentous clubbed or knobbed antennae. When at rest they hold their wings above the body with the upper surfaces of opposite wings pressed against one another. Moths on the other hand, are usually thought of as night-flying with broader or even feathery antennae. When at rest, they fold their wings, under surface down, covering the body. While there are no nocturnally active butterflies, quite a large number of moths are active during the day and some of these bear a remarkable resemblance to butterflies.

some moths that are actually more closely related to the butterflies than they are to other moths.

The Lepidoptera vary in size between tiny moths with wingspans of only 3mm to the largest with a wingspan up to 300mm. They have large compound eyes and ocelli are often present. The antennae are usually long and multisegmented and the mouthparts form a proboscis used for sucking nectar and other fluid foods. One family of moths, however, the Micropterigidae still retains chewing mouthparts and it is considered by some authorities that they should be in their own separate order. The wings are large and membranous and covered in overlapping scales, which may also be found over the rest of the body. In clearwing moths and butterflies, large areas of the wings lack scales and show the underlying clear membrane. They usually have three pairs of walking legs though the first pair may be reduced to a stump.

The lepidopteran larva has a heavily sclerotized head capsule with well-developed mandibles for chewing its food and six stemmata on each side. The head also has a pair of spinnerets from which silk is extruded. It has three pairs of thoracic walking legs and on the ten-segmented abdomen are a number of so-called prolegs. These are usually to

The most obvious feature of the order Mecoptera is the way in which the chewing mouthparts are elongated to form a rostrum and this is very evident on this male scorpion fly, **Panorpa communis**. The name scorpion fly for this insect comes from the way in which the male holds his enlarged genitalia curved upwards in the manner of a scorpion's sting.

be found on segments II to VI and X though they may actually be reduced in number in some species. The pupa is often, though not always, inside a cocoon spun from silk produced by the larva, which also produces the silk pad and girdle in those butterflies which have a naked pupa.

ORDER: MECOPTERA (SCORPIONFLIES AND HANGINGFLIES)

The Mecoptera is one of the smaller insect orders with only around 500 species so far described. They have long, slim antennae, large compound eyes and ocelli are usually present. Most noticeable about the head, however, is the 'beak'. This is a downwards projection of the head, at the end of which are the chewing mouthparts. The thorax bears two pairs of slim, similar sized membranous wings, which often have dark spots or bands on them, and six walking legs. In the hangingflies, the third pair of legs is raptorial and they are used for snatching up insect prey. Cerci are present on the end of the abdomen. In male scorpionflies, the end of the abdomen with the genitalia is swollen and held curved up over the body, giving the appearance of a scorpion tail, thus the common name given to these insects. Though the males appear to have a sting, they are in fact completely harmless. While adult hangingflies are predators, scorpionflies are scavengers, feeding on almost any kind of dead animals, including mammals and birds, as long as the internal tissues are exposed. They will also enter spider webs and feed upon prey hanging therein.

The larvae, which normally live in moist litter, are very similar to a lepidopteran caterpillar, with chewing mouthparts, three pairs of walking legs on the thorax and prolegs on some abdominal segments. Larvae of one family have no stemmata at all, while up to seven stemmata are found in other families. Unusually, some scorpionfly larvae have true compound eyes made up of 25–45 ommatidia.

ORDER: SIPHONAPTERA (FLEAS)

Despite the almost total disappearance of the human flea, at least in most developed countries, some species, notably the cat flea, may reach plague proportions in our homes under ideal conditions (for the fleas!). All fleas are parasites, feeding upon the blood of their mammal and bird hosts. Many of the 2500 described species are restricted to a single host while others, the rat flea for example, are able to live on more than one host. Fleas are laterally flattened insects, which allows them to effectively 'swim' amongst the fur and feathers of their hosts. The

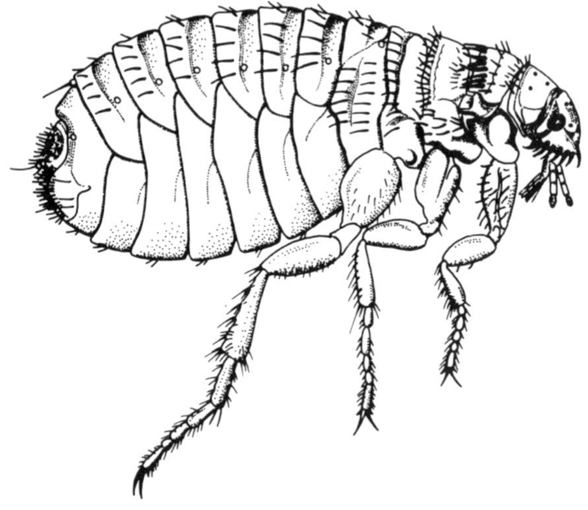

Flea.

THE BLACK DEATH

Bubonic Plague, or as it was named the Black Death, was a great killer in Europe during the fourteenth century and is caused by the bacterium *Pasteurella pestis*. It is spread by rat fleas, particularly *Xenopsylla cheopsis* and *Nosopsyllus fasciatus*. The Plague bacterium is maintained in the wild rodent population and is passed on when a flea which has bitten an infected rodent then bites a human. Bubonic Plague still occasionally breaks out today and on a recent trip to the south-western United States one of the authors was frequently warned of the dangers of getting too close to the wild rodents in the area, many of which act as reservoirs for the disease.

antennae are very short and lie in a groove on the head, compound eyes are absent but ocelli may be present. Clearly, with such a life-style, the need for any form of vision is minimal. The mouthparts are adapted for piercing the host's skin and then sucking its blood; in fleas, both sexes are bloodsuckers.

Wings, which would hinder the parasite, are absent but the flea has three pairs of walking legs, the third pair stout and adapted for jumping, so that they can move from host to host. The legs have strong terminal claws with which the flea clings onto its host. Flea larvae, unlike the parents, are mainly free-living, eyeless, legless creatures, feeding on dead skin and dried blood derived from the host.

ORDER: *DIPTERA* (FLIES, CRANEFLIES, MOSQUITOES, GNATS)

This is one of the largest insect orders with around 90,000 known species and many more probably to be described. Flies or their larvae exist in almost every habitat on Earth, with the exception of the oceans and polar ice-caps. Flies are perhaps best known for their notoriety as spreaders of disease or pests, though by no means all of them are villains.

The flies are most easily separated from other insect groups by the fact that the second pair of wings is reduced to a pair of club-shaped structures called halteres. These are used as balancing organs and contribute to the amazing aerial agility of most flies. The antennae vary in length, being longest in the more primitive families and very short in the more advanced flies. The compound eyes are well developed, often occupying a large proportion of the head surface area. In general, the compound eyes of the males are larger than those of the females and they may be holoptic, with hardly any separation between right and left eyes. Between the compound eyes are three ocelli. The mouthparts are adapted for piercing and sucking or for lapping liquid food. Since only the first pair of wings are used for flying, both the first and third thoracic segments are much smaller than the second.

Fly larvae lack true legs and in only a few families is the head fully developed. At the other extreme, in the maggot larva of higher flies, the head is vestigial with only the reduced mouthparts visible to show which is the head end. Fly larvae of one species or another feed upon just about anything which is organic or had an organic origin.

04 Sexual Behaviour

Sexual competition dominates the lives of most male insects. They have not overcome the considerable problems in reaching adulthood to play golf, watch television, go to football matches or have careers in advertising. They exist solely for a single dominating purpose – to pass on their genes to another generation. If they fail in this, they might as well never have existed. It is therefore not surprising that male insects resort to a multitude of tactics to attain their reproductive goals. If foul-play is likely to yield an adequate pay off, then it figures in the game plan, which includes deceit, dirty tricks, occasional rape and even transvestite behaviour. Playing the role of part-time 'drag-queen' can be advantageous for the males of *Cotesia (Apanteles) rubecula*, a tiny parasitic wasp. A male who is mounted aboard his mate, or has just dismounted, is faced with the problem of repelling attacks by rival males. An attractive option is to dupe a challenger by adopting a female-like 'come-and-get-me' posture, crouching low with head-bowed. With such a purely visual cue acting as an automatic releaser for the act of mounting, the challenger goes ahead and wastes time in a futile attempt to mate with the 'drag queen', leaving the latter's mate in peace. Such tactics are fair play in such a cut-throat world, and are certainly preferable to getting into a fight. This could be damaging to both participants, with

The intense competition for available females, and their frequent reluctance to mate, often leads to examples of troilism in insects. Here two males of the net-winged beetle, **Lycus constrictus** (Lycidae) from Kenya are attempting to couple with the same female. Despite her inferior size (a common trait in lycids, in which size varies greatly in both sexes) she is refusing to accept either of them.

possible long-term costs through physical injury, and the highly likely short-term cost that the female will wander off and mate with another male while the two rivals are too busy laying into one another to notice that the female has gone.

Males thus devote more time and energy to the sexual chase than to any other activity, but in doing so may harass the females to an extent which is not in their interests. Many females mate only once, or at irregular intervals, and need to devote most of their time to safeguarding the future for their offspring. Being constantly pestered by overzealous males can seriously interfere with this work, and even interrupt it for lengthy periods if prolonged matings are imposed. Females have evolved various measures for trying to reduce male nuisance-attacks, such as raising the abdomen in a 'not today dear' gesture (for example many butterflies, assassin flies, damselflies). In many large dragonflies, the females go about the business of egg-laying in an inconspicuous manner, helped by their coloration, which is cryptic, in contrast to the bright colours of the males. In the ground-beetle, *Pterostichus leucoblandus*, the female lays an unwanted suitor low by blasting the male full in the face with a spray directed from the tip of the body. It takes him anything up to 3 hours to wake up from this treatment, by which time the object of the male's abruptly terminated attentions will be long gone.

A life tormented by pushy oversexed males is not the norm for all female insects, simply because in some species the male sex just does not exist. This is especially so in many tiny parasitic wasps which develop inside the eggs of other insects, with generation after generation being churned out asexually through a process known as parthenogenesis. This is also the normal method of reproduction during the summer months in aphids. It is the all-

The inferior stature of male grasshoppers compared with their mates is obvious in this pair of **Auloserpusia phoeniconota** mating in a Kenyan rainforest.

out ability of a brood mother to mass produce a continuous succession of live-born exclusively female offspring which enables these tiny insects to reach the astronomical numbers needed to beat the odds in the face of their numerous adversaries. However, when autumn arrives the baby-machine winds down and normal sexual reproduction reappears. Males are suddenly produced, mating takes place and amazingly large eggs are laid which can survive the rigours of winter.

FINDING A MATE

With insects being generally rather small, and often widely scattered within a given environment, the problems posed by actually finding one another would seem to be a major stumbling block facing productive sexual liaisons. Given that several hundred other kinds of insects are also going to be busily engaged in solving the same problem in the same living space, there is a considerable potential for confusion, due to the amount of 'noise' around, interfering with individual sexual communication. Sorting out the correct signals from the background 'clutter' can be achieved through a number of avenues, some of which work well at long range, others only close up. One common ploy is to switch from one system, such as sound, for long-range communication, to another, such as touch, at close range.

The two methods which excel in bringing the sexes together from a considerable distance are sound and scent. Light will also work well when employed after dark, but normal visual recognition in daylight only works well at fairly close range. However, various tactics can increase the chances of getting a good view of one another without having to undertake vast amounts of time-consuming searching. Male insects, which appear to be merely sitting around idling away their day, may therefore actually be engaged in the deadly serious business of getting a mate. For a male carder-bee this may involve spending all day defending a particularly rewarding patch of flowers against both rivals and other flower-visiting insects. Females are guaranteed to put in an appearance sooner or later at such a high-quality source of nectar and pollen, both of which have been conserved for them in superabundance via the exclusion zone imposed by the diligent male. The male's reward, in sexual terms, is mating with any arriving females, with the extra pay-off that there will be plenty of food supplies around for provisioning nests containing future offspring.

Males of many insects practise such resource defence, providing a predictable sexual rendezvous as long as the defended resource (which could be a sap-run on a tree or a patch of mud as well as flowers) is of sufficient value to be virtually certain of attracting females. Neither should it be so common that it is highly unlikely that a female will happen to show up at that particular site. The system works well for resources which are always in short supply or distributed in discrete patches, such as dung or small animal corpses.

The need to expropriate and then hang on to such resources against allcomers has produced some impressive fighting equipment in certain male insects. The conspicuous horns or enlarged mandibles of rhinoceros and stag beetles are particularly striking examples of highly developed duelling pieces. The Central American weevil, *Macromerus bicinctus*, engages in ritualized 'club fights' using its enlarged front legs as cudgels for belabouring an opponent for possession of squatter's rights on a fallen log. The grotesque males of the Central American brenthid beetle, *Brenthus anchorago*, use their extravagantly elongated snouts as lances to side-swipe one another off a tree, the moribund condition of which will attract females for egg-laying purposes. This long snout is also useful for laying across a female as a barrier, while the female methodically drills into the wood with a much shorter snout prior to laying an egg. By guarding the female in this way the male keeps away rivals and is able to ensure that the egg will be fertilized by himself, delaying the moment of mating until the female stops drilling and turns to lay an egg. Similar habits are found in other equally bizarre brenthids around the world.

SOUND

To a human observer of insect activity, sound is the most obvious form of sexual communication, although it may not be obvious that the (to us) unwelcome whine of a female mosquito's wings generates a species-specific frequency detected by the Johnston's organs on the male's antennae. By contrast, the stridulatory 'love songs' of leafhopper bugs or the ultrasound clicks produced by the microtymbals of some moths are outside the range of normal human hearing. This is certainly not the case with most cicadas, in which the volume and quality of sound produced by even a single singer can be painful to the human ear at close range. Many species aggregate in single trees, creating a sound beacon audible over a huge area and deafening for a human listener. We do not know how females 'choose' an individual caller for a mate from within such a cacophony. In a few species the male cicada flies around and utters a croaking call, rather than singing from a perch. Some examples of insects using sound for sexual communication are shown in the box.

The most familiar insect sound to those living in cooler northern climes is the chirping of grasshoppers. In most species this is produced solely by the males violin-style by rubbing a 'file' on the inner side of the hind legs against a 'scraper' on the adjacent forewing. Occasionally there is also a more subdued female-response song. Stridulation alone is by no means adequate to lead to mating, but merely brings the sexes together, after which the suitor must perform the dance as well as the song. Courtship routines may call into play a considerable repertoire of visual displays using the legs, antennae

INSECTS USING SOUND FOR SEXUAL COMMUNICATION

SPECIES	TYPE OF SOUND
Gromphadorhina portentosa, cockroach (Madagascar)	Hisses in a close-up courtship context by expelling air through the second pair of abdominal spiracles
Isoperla grammatica, stonefly (and other Plecoptera)	Drums abdomen rapidly against substrate
Nezara viridula, green vegetable bug	Males stridulate, producing seven distinct songs, including a duet with rival males; females have three songs
Buenoa spp., backswimmer bugs	The male stridulates by rubbing special roughened areas of the front legs against a stridulatory zone at the rostrum's base
Xestobium rufovillosum, death-watch beetle	The male bangs his head against the wood on which he is standing
Psammodes spp., tok-tokkie beetles	The male rapidly taps the underside of his abdomen against the ground
Drosophila, Zaprionus and *Scaptomyza*, fruitflies	Males of most of these and other fruit-flies produce 'love songs' by rapid vibrations of their wings
Galleria mellonella, greater wax moth	The male generates brief ultrasonic bursts from small tymbals as he flutters his wings
Syntonarchia iriastis, Australian pyralid moth	Male produces a torrent of ultrasound using a file-and-scraper device on the genitalia and abdomen
Hamadryas spp., cracker butterflies	Males produce a loud snapping sound as they fly

and body. Among the more accomplished performers is the European mottled grasshopper, *Myrmeleotettix maculatus*. Yet in some other species it is the male's ability to stick tight on a wildly bucking female, following a rapid and guileless jump-and-mount approach, which decides whether or not he will eventually be accepted.

In common with most insects, male grasshoppers are quite a lot smaller than their mates, which adds to the difficulty of staying put once mounted. Exceptions to this 'small male' rule include mutillid wasps, in which the winged males greatly outweigh the tiny wingless females, who are literally whisked off their feet and air-lifted to some safe place where mating can proceed without interruption.

Katydids and crickets stridulate by rubbing together the specially adapted forewings in a blurr of movement. In mole-crickets this produces a continuous churr of sound with a powerful ventriloqual element, making it very difficult to locate the singer. The sound is amplified by the caller sitting in the mouth of a funnel-shaped burrow, which differs in shape according to species and the wavelength of the song. Some crickets also boost their volume by calling from sound systems built of mud. In a few *Oecanthus* tree-crickets, the male perches in a pear-shaped hole which he chews in a leaf, this thereby acting as a sounding-board. In most crickets, the louder the song, the more attractive the singer to an approaching female, although she may well be waylaid close to the caller's burrow by a 'satellite' male or 'sneak'. Poor-in voice himself, he relies on a strategy of silence, stealthily lying in wait in the (usually fruitless) hope of stealing a quick illicit mating with an in-the-mood female lured by a stentorian-voiced rival. Yet being two-timed by a 'sneak', though bad in itself, can be the least of the troubles suffered by a loud-voiced male cricket. Females of certain tachinid flies home in on the singing males, especially the loudest ones. The flies then lay their eggs upon them, leading to an eventual unpleasant death as they are eaten alive from within by the flies' larvae; it is the mute 'sneaks' who are most likely to escape such a fate.

A rather more abrupt death may overtake male tropical katydids as they stand and transmit their nocturnal calls from a rainforest leaf. These reach the ears not only of the intended females, but also of listeners-in, such as katydid-eating bats, which use the songs to indicate the precise location of their next meal. Some callers, such as the Neotropical unicorn katydid *Copiphora rhinoceros*, try and minimize the risk by singing in short bursts sufficient to attract females, and then switch to a less risky close-up tactic called tremulation when one arrives. A tremulating male communicates to the female by vibrating the leaf on which he is sitting, a rather less public method than singing.

PHEROMONES

Sound therefore imposes certain implicit risks as a sexual communications medium, for the simple reason that the caller is broadcasting to potential predators within earshot. Scent can be much more species-specific, and has the advantage of carrying even further than sound, such that the sexual scent

By fluttering her wings and simultaneously releasing a pheromone from her abdomen, this female **Odontotermes** sp. termite will quickly attract a mate in the immediate aftermath of a thundershower in a Nepalese forest.

(known as a pheromone) produced by certain female moths can be detected by the males at distances of many kilometres – way beyond the range of even the loudest sound broadcast. The female moth may not release the pheromone in a steady stream, but in a pulsed action, as seen in the American arctiid, *Utetheisa ornatrix.* However, in some arctiids, such as the American saltmarsh moth, *Estigmene acrea,* it is groups of males who dispense their scent messages from huge bladders called coremata, which can be inflated at the rear of the abdomen.

The receptors situated on the many-branched pectinate antennae of the males are capable of responding to only a few molecules of the appropriate female pheromone, but are indifferent to the scent of another species. Being mainly nocturnal, moths are heavily reliant on pheromonal communication (although a few also stridulate), although scent also plays a 'come-and-get-me' role in the females of certain butterflies such as the European silver-washed fritillary, *Argynnis paphia.* In some neotropical heliconiine butterflies, even the pheromones released by the female pupae are enough to gather bevies of males around them. Such is the compulsion to be first at the sexual checkout that attempted rape of the unhatched pupae is not uncommon. However, in most butterflies scent mainly comes into play during normal close-up courtship.

The release of pheromones as a long-distance attractant, known as 'calling' behaviour, is also seen in male fruitflies of the family Tephritidae. In the Mexican fruitfly, *Anastrepha ludens,* the diminutive male increases the range of his output by raising his abdomen and then inflating two pouches situated on either side of it. These are the pheromone generators, and he spends many hours perched beneath his citrus leaf calling

It is not only male moths which possess pectinate antennae for long-range detection of pheromones released by females. Many beetles also boast similar devices, including this **Calyptocephalus gratiosus** (Lampyridae) from Trinidad.

INSECTS WHICH USE SCENT AS A LONG-DISTANCE ATTRACTANT

♦ *Latiblatella angustifrons*, Honduran cockroach. The female adopts a 'calling' stance with drooped abdomen and raised wings. 'Calling' behaviour is known in three of the five families of cockroaches.

♦ *Eurygaster integriceps*, stinkbug. The male 'calls' by bobbing the upraised abdomen up and down in a pumping movement.

♦ *Nezara viridula*, green stinkbug males call by releasing a pheromone which attracts males and females for mating aggregations.

♦ *Climaciella brunnea*, mantispid males release a volatile pheromone powerful enough sometimes to attract bevies of females.

♦ *Bittacus apicalis*, hangingfly. The male attracts a female to sample the insect gift he has caught by hanging from a leaf, shivering his wings and bending his abdomen upwards to evert pheromone-releasing abdominal vesicles.

♦ *Physiphora demandata*, otitid fly males 'call' by releasing a pheromone from a blob of fluid exposed on the extruded anal gland.

♦ *Hemithynnus hyalinatus*, thynnine wasp females in Australia 'call' for a mate from just beneath the surface of the ground. If a male fails to turn up within a day or so, the female climbs up a plant stem to increase the scent's dispersal range.

platform sitting thus. The inflation of abdominal pouches is known in many other species and in the chloropid fly, *Thaumatomyia notata*.

Much has been made of the supposed grisly habits displayed by female praying mantids in chewing off their mate's head while his rear end continues to carry out its procreative duty. Yet such sexual cannibalism is probably quite rare in nature, mainly because in at least some (and probably many) mantids the female actively solicits the males to come and pay a visit by 'calling'. In the neotropical mantis, *Acanthops falcata*, the female, who looks extraordinarily like a crinkled leaf which has been dead for a very long time, hangs beneath a twig, bends her abdomen downwards and, as dawn steals across the darkened rainforest, sows her pheromonal message on the humid air. Upon answering such an emphatic request for their inseminatory services, the males are assured of a safe-sex encounter when they arrive.

Termites engage in a simultaneous mass exodus of males and females from most of the nests within a given area. Unlike the worker termites, these prospective 'kings and queens' are fully winged, earning them the name of alates. After a short flight from the vicinity of her nest, the female *Armitermes euamignatus* from Brazil sheds her wings (snapping them off easily at special basal fracture zones) sticks her abdomen skywards and pumps out a pheromone to 'call in' a male. In some species 'calling' takes place before shedding the wings, which can thereby serve one last extra purpose by being flapped vigorously to help disperse the scent more widely; in *Hodotermes mossambicus* from Africa, it is the male who 'calls'. Many female ants do likewise, climbing a stem near the nest and exuding a droplet of highly volatile sexual fluid from the sting. In many American *Pogonomyrmex* harvester ants, it is massed ranks of males who release a veritable plume of scent from their mandibles. This gradually seeps downwind and attracts yet more males to form vast mating assemblages. Most remarkably, these form on specific sites, often of no observably special nature, which are used regularly year after year by ants who have never seen the place before.

LIGHT

Light acts as a long-distance lure for relatively few types of insects, but is especially important in the firefly beetles of the family Lampyridae, as well as in some other beetle families. Fireflies (also called glow-worms) are concentrated in warmer countries, there being but a single species in the British Isles; the USA boasts a rich variety. In many species it is the females who sit motionless on vegetation and switch on their miniature sexual lighthouse beacons to lure in passing males. In other species, it is the males who issue the nocturnal sexual invitations, flying around and emitting either a continuous very bright light or else a series of flashing pulses.

With many different species of fireflies often being on the wing simultaneously in the same patch of forest, any waiting female is seemingly faced with a problem in knowing which male to respond to by switching on her own light. Chaos is avoided by each species flying at a different level in the forest, or at a different time, while also using a different 'code' of flashes, varying in colour, flash length, flash interval and the number of flashes in a given sequence. In the Asian genus *Pteroptyx*, an individual display gives way to a massed light-fantastic as millions of males gather in certain trees, to flash the night away in a synchronized display of startling intensity. Unlike an array of massed light-bulbs, such a display of photonic firepower does not release any excess heat, as light production in fireflies is a highly efficient biochemical process, during which luciferin is oxidized by an enzyme luciferase to produce a 'cold' light.

BIOCHEMISTRY OF LIGHT PRODUCTION

The production of light by insects has been studied in most detail in the lampyrid beetles, most commonly known as fireflies. Light production, at least in most artificial systems invented by man, also involves a considerable accompanying heat output, but this is not the case in insects, who produce cold light in a carefully controlled series of reactions. Both the rate of flashing and the colour of the light emitted can be varied, the latter by altering the pH of the reactants. The essential part of the reaction, which releases the light plus carbon dioxide as a waste product, is the oxidation of luciferin to oxy-luciferin by the enzyme luciferase. Like most biochemical processes, energy has to be put into the reaction in the form of ATP (adenosine triphosphate). While the lampyrids, some other beetles and a few other insects are self-luminescent, others do so with the cooperation of symbiotic light-emitting bacteria or fungi.

RENDEVOUZ POINTS (LEKS)

If sound, scent or light are not available, the two sexes must find some other way of locating one another. One common method is to form swarms around easily recognized markers, such as a lone tree or bush, a puddle or a sunlit spot over a river. Even the top of a human head may provide a temporary outstanding marker. Swarms tend to be used by insects which hatch widely and unpredictably over a broad area. These include mosquitoes, gnats and many other kinds of flies, as well as many termites. Swarms tend to be composed solely of males, with females arriving for just long enough to find the only mate they will ever take. In most swarming insects, the males are easily distinguished by their extremely large holoptic eyes, meeting at the top of the head, and designed to enable quick recognition of any female who joins the swarm. In male mayflies (Ephemeroptera), the ommatidia in the upper part of the eye are extra large, or may even form a separate upper eye, enabling the male to spot females coming down from above as he flutters downwards within the swarm.

Many male solitary bees are faced with the problem of locating females emerging from nests which are widely dispersed, making it unlikely that a mate can be found at the actual nest site without expending a vast amount of unprofitable energy in searching. If the female is of a species which specializes on a single type of flower, than an open bloom makes a good rendezvous point, as for example in the American *Perdita opuntiae.* The males lay sexual ambushes for females on the prickly-pear cactus flowers which are the only ones they ever visit.

A mating pair of common darter dragonflies, **Sympetrum striolatum**, in England. The males are territorial around still water and normally remain in tandem with the egg-laying female.

If the flower species is clumped, then the males can defend the whole clump against other contenders until a female shows up. Yet one species of *Anthophora* bee from the Sinai desert in Israel devotes many hours each day to a wearisome patrol around 'his' clump of flowers, despite the fact that during this period 'his' coterie of females are not actually calling in for supplies of pollen or nectar. However, there is just enough overlap each day for a single mating, and this is probably sufficient to ensure that the flower-owning male fertilizes any egg laid that day. Unlike in many such territory-owning bees, he is in a position to check out what is happening back at the nests, as these normally occur within his flower-holding territory.

In some instances males and females really seem to come up against an insurmountable problem in arranging a liaison. Take the case of the hoverfly, *Volucella bombylans*, which is widespread in Europe and North America. The larval life is spent as a scavenger within the nests of bumble bees, to which the adults bear a striking resemblance. Bumble bee nests are widely and unpredictably dispersed, making them economically unfeasible as sexual rendezvous points. Unlike in some bees, the females of *V. bombylans* roam widely and feed on a broad selection of flowers. This renders searching on flowers a rather thankless proposition, the more so as the females so closely resemble bumble bees that any questing male is misguided into jumping on the wrong target. The answer, as with many insects faced with similar problems, is to form a lek, a rendevouz point, in which several males compete for small individual territories within a locality which can be recognized by the female. She will head to such a place specifically to find a mate, and for no other reason, as lekking sites do not normally hold any resources of special interest to the females. In *V. bombylans*, the males usually space themselves a meter or two apart on leaves fairly close to the ground along a warm south-facing hedgerow or woodland margin. They fly up to meet and briefly contact any likely looking insect which passes overhead. When a female puts in an appearance, it is noticeable how quickly she accepts one of the males as a mate, without the 'play-hard-to-get' struggle so typical of most female insects when suddenly and unceremoniously grabbed by a suitor. This is ample confirmation that she has indeed arrived for just a single purpose – sex. Lekking is found in many other insects, from butterflies to robberflies and hunting wasps, and seems to be particularly common in species which inhabit deserts. Some examples are shown in the table overleaf.

EXAMPLES OF INSECTS THAT FORM LEKS

SPECIES	TYPE	LOCALITY
Orchelimum spp.	meadow katydid	low vegetation, North America
Tabanus bishoppi	horsefly	forest, North America
Hylemya alcathoe	anthomyiid fly	on low leaves in woodland, North America
Hermetia comstocki	soldier fly (Stratiomyiidae)	on Agave plants, Arizona Desert
Lordotus pulchrissimus	bee fly (Bombyliidae)	California Desert
Hepialus humuli	ghost moth	meadows, Europe
Atlides halesus	great purple hairstreak butterfly	desert ridgetops, USA
Hemipepsis ustulata	tarantula hawk	on palo verde trees on ridge tops, Arizona Desert.
Polistes canadensis	paper wasp	ridgetops in dry forest, Costa Rica
Xylocopa varipuncta	carpenter bee	ridgetops and dry washes, Arizona Desert

In many insects, males search out females on flowers or at nest sites, often competing with one another in a scramble to attain sexual conquest. Success rates are often depressingly low, forcing the males to try even harder, so it is common to see male beetles or tephritid flies jumping onto the backs of several females in succession, only to be firmly rebuffed by being shaken off or thrust away with the back legs. In a number of bees and wasps which nest in dense aggregations in the ground, the males become sexual groupies, clustering over nest holes containing emerging females and frantically competing to be first to dig them out and

Several **Tetralonia malvae** males become covered in dust as they fight over a newly emerged female in southern France. Such behaviour is found in a number of bees which nest in dense colonies, and is possible because in most bees the males emerge before the females.

A male green tiger-beetle, **Cicindela campestris** (Carabidae) clasps his jaws securely around his mate as she probes deeply with her ovipositor into a coastal sand-dune in North Wales. By guarding her in this way, he makes it difficult for a rival to steal the female at the vital last moment, just before she lays her eggs.

claim them as a mate. The usual result of such intense rivalry is the formation of 'mating balls', as scuffling males coalesce around the body of the unfortunate virgin.

With such intensity of competition being the norm, there is an enormous premium for a male being able to hang on to his mate in the face of a barrage of efforts from his rivals to unseat him. In many beetles, particularly scarabs, the rider's struggles to stay mounted are greatly aided by his hold-fast feet, long up-curving tarsal claws which can be hooked securely under the front of the female's thorax and legs. Male tiger beetles (Carabidae: Cicindelinae) clamp their broad powerful jaws around a corresponding groove (coupling sulcus) on the female's thorax. This snug fit thwarts both the efforts of rivals or indeed the recalcitrant female herself to unseat him.

A pair of **Volucella bombylans** hoverflies mating in an English garden. The way in which the male hangs downwards suspended by his genital claspers, rather than riding atop the female's back, is typical for this species.

Competition for females on temporary 'bonanza' resources, such as dung, are particularly intense. In the numerous takeover attempts, which frequently spill over into full-blown fights, in the common yellow dungfly, *Scathophaga stercoraria*, the female herself may become damaged or even killed, forced face down into the mire of the cowpat, mortally wounded by the powerful kicking spiny legs of her would-be suitors. Not surprisingly, females coming to the pat to lay their eggs prefer the largest mate they can acquire, best able to defend her should a rough-and-tumble develop. Males of the tiny black sepsid fly, *Sepsis cynipsea*, which often inhabit the same cowpats, are less prone to dethronement from their mates, having a handy 'stirrup' available, a spine-and-notch lock on the forelegs which engages securely with the female's wings. Males of the African elegant grasshopper, *Zonocerus elegans*, have an especially large investment to lose if they are pulled off their mates by a burly rival. In this species, the male stays on the female's back for weeks as a rider, mating regularly and transferring numerous small spermatophores to the female. She probably uses these to help nourish her developing eggs. When the time comes to lay her eggs – after fresh rains – numerous pairs assemble at favoured egg-laying sites. There they are faced with a salvo of assaults from desperate bachelors, who effectively blockade the egg-laying sites in a last-ditch gambit to avoid the prospect of a sexual lockout. This is make-or-break time for the male riders, who stand to lose their entire investment in spermatophores at a single stroke. This is because whichever male manages to mate with the female just before she oviposits, will father most of her eggs. Fortunately, the riders secure a firm grip with their front legs and are difficult to pull off their mounts, peddling their brightly coloured back legs as a preliminary warning that a powerful kick in the face is in prospect for any contender for their mate.

In many lygaeid bugs, such as these **Lygaeus reclivatus** from Mexico, the male remains connected to the female for as long as possible, sometimes up to 24 hours or more. Such prolonged mating times deny access to the female genitalia until shortly before she lays her eggs, thus constituting a form of female-guarding behaviour by the male. Like most insects which feed on milkweeds (Asclepiadaceae), this species is warningly coloured.

COURTSHIP AND MATING

JUMP-AND-HOPE TACTIC

Courtship finesse is often scant or totally lacking in insects – the males merely adopt a 'jump-and-hope' tactic by mounting the females without any preliminaries. The strategy which follows is one of sheer persistence, trying to wear down the female's resistance until she finally gives in. This does not always work, as the female often has a few tricks for thwarting even the most strenuous efforts of a resolute suitor. The most obvious ploy is to deny access to her genitalia, either through refusing to open the necessary genital pore, or perhaps by bending her genitalia down out of reach of the male's genital claspers. Even after he has connected his genitalia correctly, there may still be internal female barriers which prevent the sperm reaching the correct place. She can also throw the male off by shaking violently, or even round on him and mount a vicious attack, perhaps administering some unwelcome 'love-bites' by chewing off a few legs or bits of his antennae before he manages to make good his getaway. Such rough behaviour is common in some longhorn beetles.

The full gamut of courtship procedures, from jump-and-mount to protracted and complex routines, can be well illustrated in a single group of insects such as the robberflies (Asilidae). In most species, the males simply drop in on a perched female and try to force genital contact. In the American *Efferia varipes*, the female may try to fool an unwanted suitor by playing dead until he loses interest in trying to mate with a 'corpse' and flies off.

DANCE

A more sophisticated approach employed by many males is to perform some kind of courtship 'dance', either while perched in front of the female or by hovering in front of her while she is perched. Some of these performances are quite elaborate, especially in the various species of *Cyrtopogon*.

COURTSHIP DANCING IN *CYRTOPOGON* FLIES

In these handsome flies, the front legs and ornate golden-haired abdomen are waved and bobbed right before the female's eyes in what is supposed to be a mesmerizing piece of ballet. It is therefore puzzling why success rates are in fact abysmally low, with the females usually either ignoring the males' dramatic efforts, or simply flying off to find somewhere more peaceful. More is said on the dance-flies on page 81.

FEMALE COOPERATION

In dragonflies and damselflies (odonates) courtship is generally absent, although a high degree of female cooperation is always essential because of the peculiar stance adopted during the act of copulation. Before he can actually mate, the male has to transfer semen from the genitalia situated near the tip of his abdomen, to a set of accessory genitalia way up near its base. Given the flexibility of the slender abdomen, this can be achieved directly by bending the abdomen forwards and upwards, an act which may take place before or after the male has contacted a female, depending on species. Many male odonates are territorial over water, a gambit which is bound to pay off sooner or later as females simply have to show up eventually in order to lay their eggs. When she does, she will be grabbed by the nearest male who will attempt to grasp her with his terminal claspers. These are differently shaped according to species, and designed to make a snug fit behind the female's head (in dragonflies) or around the front of the thorax (in damselflies). Once in the tandem position, male leading, female clamped on at the rear, the next move is up to her. She must bend her abdomen inwards and upwards, perhaps making a few ineffectual stabs at locating the spot where their two sets of genitalia can connect, before she at last succeeds. Any show of reluctance on her part, and she may be encouraged to perform with a brisk wing-clap from the male as he lifts her upwards. Males are usually over anxious to bag almost any female which arrives, and often get hold of the wrong species. This rarely, if ever, leads to mating, as the 'feel' of the male's claspers around her will inform the female that she has been hijacked by the wrong male, and she will not receive the correct stimulus to raise her abdomen to the coupling position. Eventually the male releases her.

WINGS AS SIGNALS

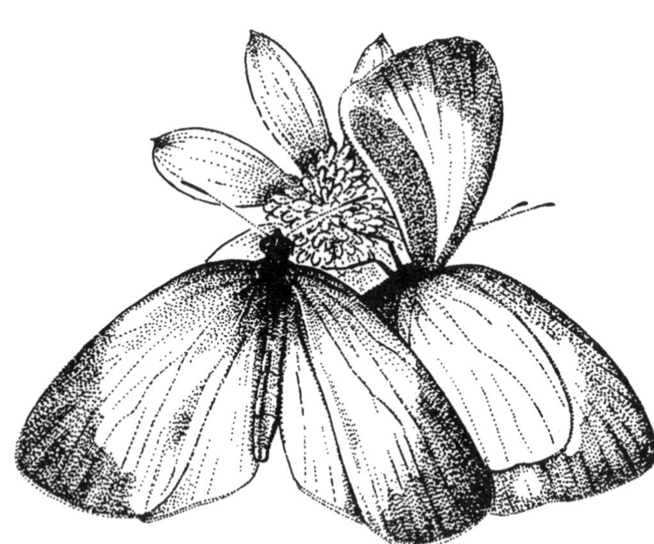

In the American barred sulphur **Eurema daira** (Pieridae), the male (right) sits beside the female and performs a 'wing wave'. While starting out as a static salute, it soon develops into active courtship as the male sweeps the wing up and down beside the female's head. The initial signal is purely visual – the wing being waved bears a black band on a yellow background, a colour absent in the female. However, during the active waving courtship the wing's trailing edge brushes the female's antennae and probably passes on a contact pheromone.

The courtship of the fly **Poecilobothrus nobilitatus** (Dolichopodidae) is a hectic affair in which the male (right) vigorously scissors his wings in front of the female. Courtship usually takes place on the surface of shallow water, such as a temporary puddle. Unlike the female's plain wings, those of the male are tipped with white, creating a conspicuous blurred effect during courtship, which usually enjoys a spectacularly low rate of success!

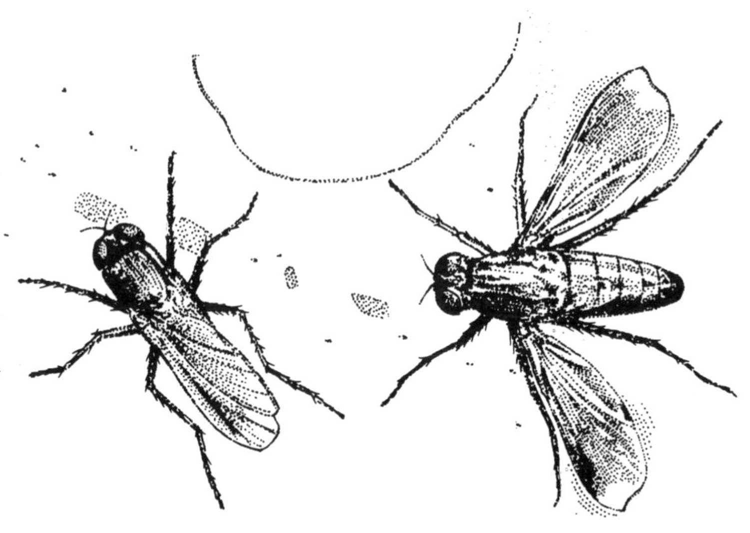

SIGNALLING

Some damselflies have evolved a most complex repertoire of courtship signals, probably the most complex known in the invertebrate kingdom, and far beyond what most people expect insects to be capable of originating. This is especially so in the dazzlingly metallic members of the genus *Calopteryx*, commonly seen flaunting their flashy livery in cavorting flights over running water in Europe and North America. In the European banded demoiselle, *Calopteryx splendens*, the males fight long hectic helter-skelter aerial battles to establish a territory around a clump of waterplants. The disputed territory has been selected for its qualities as a potential egg-laying site, and the males are equally as skilled as the females in choosing the best spots. In fact the better the site, the harder the battle for its possession. When a female comes flipping over the water towards him, the territory-owner belly-flops off his leaf into the water. He tries to land right above the precise spot he has chosen as the prime one for egg-laying. He floats there with his wings held in a special 'X-marks-the-spot' configuration. If the female accepts his invitation, she makes a zig-zag pass overhead and then flies to a nearby stem, beside which the male performs a wing-whirring flight, confirming who he is, and they eventually mate. She then follows him back to his territory, often showing a staggering ability to remember exactly where he had belly-flopped, and gradually crawls beneath the water to lay her eggs in the submerged stems of the water plants, while her mate stands guard nearby. There are many other gestures, employing both the wings and the abdomen, pressed into service by this amazing species, comprising an impressively full 'vocabulary' of signs.

COURTSHIP OF SOME INSECT SPECIES

BUTTERFLIES

Butterfly courtship usually involves a combination of stimulation via scent, touch and colour. The latter is often strikingly different in the sexes, for example in many 'blues' (Lycaenidae), in which only the males are usually blue with the females being mainly brown. Territorial behaviour is a common feature in male butterflies, as in 'hilltopping' species such as the Chilean swallowtail, *Battus polydamas*, which defends the rocky tops of small hills in the Atacama Desert, to which females come for mating. The second main strategy is patrolling, in which males such as the European orange-tip *Anthocharis cardamines* cruise a beat looking for females at rest. Females of some species, such as the European ringlet, *Aphantopus hyperanthus*, actively solicit the attention of nearby males by performing a special 'Lolita' style of 'come-and-get-me' flight. Female small heaths,

MATE REFUSAL IN PIERID BUTTERFLIES

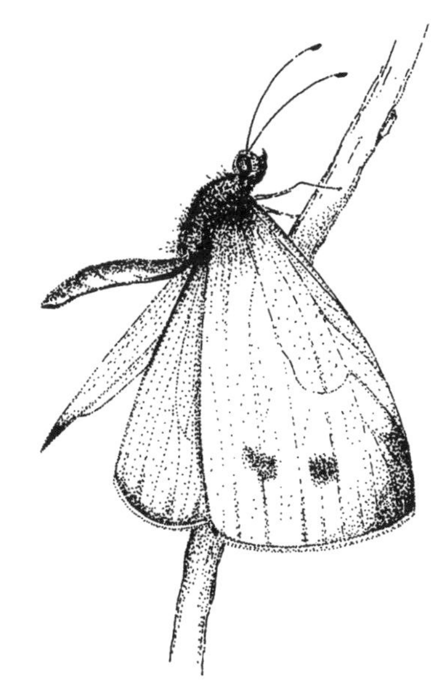

The 'not-today-dear' mate refusal posture by female pierids involves drooping the wings and raising the abdomen, which is often shivered. This seldom induces a male to desist from his courtship at once, and he often redoubles his efforts, but he usually gives up in the end and flies off to search out a more receptive female. Alternatively, the female herself tires of the male's insistent buffeting and flies off rapidly in an effort to give him the slip.

Coenonympha pamphilus, perform likewise when they enter male lekking areas.

Close-up courtship in butterflies generally involves the male flying above a female and buffeting her with his wings. He simultaneously showers her with special pheromone-impregnated scales called androconia, which become detached from his wings. Many male danaines, such as the African monarch, *Danaus chrysippus,* can evert large brush-like hair-pencils from the rear of the abdomen. Particles detached from these bombard the female with a seductive odour message. Procedures in the well-known American monarch, *D. plexippus,* are generally less sophisticated. The male simply sets off in hot pursuit of a female and tries to leg-clamp her from above in an 'aerial take-down', designed to bring her to earth so they can mate. Male danaines derive their seductive pheromones from extraneous sources, such as withered heliotrope plants, which are rich sources of chemicals called pyrrolizidine alkaloids.

In some butterflies, notably Acraeinae, Danainae and the Parnassiine group of swallowtails (Papilionidae), males ensure future sexual abstinence in their mates, at least in the short term, by plugging up their genitalia with a device called a sphragis. This 'chastity belt' is derived from the male's accessory glands, in the form of a viscous liquid which soon hardens on contact with air to form a hard plug.

In butterflies, males and females are often differently coloured. This is seldom more graphic than in **Cymothoe lurida** (Nymphalidae) photographed in a Kenyan rainforest.

Male above.
Female below.

COURTSHIP IN MOTHS

In many moths, especially in the huge family Noctuidae, males possess reversible scent-brushes situated on the legs, thorax or abdomen, although most often on the latter. These are not generally long-distance attractants, but come into use during the more intimate stages of courtship, to douse the female with scent.

BEETLES

Close-up use of pheromones is also widely used in beetles, although scent probably also plays an important role in long-distance attraction in many species. An original method of getting a female 'in-the-mood' employed by a number of male beetles is the use of so-called excitators situated on various parts of the body, usually the head. In order to bring these into play, the male must coax the female to come right up to him, so that he can introduce her to the delights of these organs. In *Malachius bipustulatus*, a small and very common European member of the Melyridae (also seen as Malachiidae), the excitators appear as two yellow excrescences situated at the base of the male's antennae. Enticing a female to take an exploratory nibble at these is easy, as most females seem only too willing to indulge in a fairly lengthy bout of sexual snack-taking, and readily approach a male head to head. Despite this, success rates in this and other related species are incredibly low. All too often, after depleting a male's excitatory supplies during a protracted nibbling session, the female makes off without showing the slightest sign of sexual arousal.

In the North American *Tegrodera*, oil-beetles, the male has to persuade the female to stop running away and face him for long enough for him to lassoo her antennae with his own. He then pulls them down into two excitatory grooves on the front of his head. The need for the male to encourage a female to feed on parts of his own body is also common in various crickets and cockroaches. In the Central American cockroach, *Latiblatella angustifrons*, the male raises his wing to expose a bristly gland on the upperside of his abdomen. If she accepts the implied invitation, the female moves forward and feeds on the glandular excitatory secretion, which in view of its purpose is called seducin. It encourages her to move right forward on the male's back and to open her genital pore so that her mate can attach a spermatophore. The lengthy session of courtship typical of the North American *Pteronemobius* crickets is designed to persuade the female to mount the male's back, from which position he can transfer a spermatophore. Given the chance, the female will immediately turn and eat this as soon as she dismounts, destroying the valuable sperm within. Therefore the male absolutely must keep her in place on his back long enough for the spermatophore's contents to make the crossing safely into the female's body. He manages this by holding his back legs at a convenient angle, so that the mounted female can feed on an attractive substance contained in hollow spines on the tibia. In another American species, *Hapithus agitator*, it is the tips of the male's raised wingcases that are presented for the female to chew away on, to such an extent that they usually become noticeably frayed and tatty at the edges. In *Cyphoderis buckelli* the male's wings are exceptionally fleshy, serving up to the female a meal not only of meat but of blood as well as it leaks from the wounds induced by her jaws. A gin-trap device at the male's rear end clamps her in place while she dines, guaranteeing ample time for spermatophore transfer.

KATYDIDS

In katydids (Tettigoniidae), the sperm-containing sac in the spermatophore is augmented by a large protein-rich fleshy appendage called a spermatophylax. This can represent quite a large proportion of the male's total body weight, some 30–40 per cent in the European tizi, *Ephippiger ephippiger*. Shortly after her mate has departed, the female katydid bends her abdomen upwards and browses on the spermatophylax which bulges prominently from the rear of her abdomen. In some species, it can take an hour or more to consume such a bulky meal. This allows a safe margin for the sperm to pass into the female's body before she has chewed her way through the outer wrapping. It may also make a nourishing contribution to her developing eggs leading to larger, healthier offspring. Males who make such bulky investments mate only every few days, as the production of such a large mass of protein takes time. The other katydid tactic is to mate frequently and serve up relatively small spermatophylaxes whose nutritional content is low. These serve almost exclusively as sperm-protection delaying devices.

Males of the North American Mormon cricket, *Anabrus simplex*, supply such a generously proportioned spermatophore that in unproductive arid habitats, where only a few males manage to wax fat, the females fight among themselves over access to mates. This is a role-reversal which contrasts with the normal trend in insects whereby males (who normally contribute nothing but a tiny amount of sperm) compete for females (who contribute a comparatively vast amount in terms of eggs). With the average male katydid making such a lavish reproductive investment, it is in his interests to ensure that his mate's eggs are fertilized by him alone. In the Australian *Requena verticalis*, the first male to mate with a female will fertilize most of any subsequent eggs, even though she may take several more mates. Each one of these would therefore be fooled into donating a large spermatophore for little or no return. To avoid this, the males are able to assess the age of their prospective partners, preferring young females who are more likely to be virgins.

NUTRITIONAL OFFERINGS

In many true flies (Diptera) the males exhibit considerable ingenuity and originality in the types of edible materials they serve up as nuptial gifts for their partners. In several kinds of *Drosophila* fruitflies the

The male of these **Dichopetala** sp. katydids (Tettigoniidae) in Mexico is just beginning to transfer to his mate the bulky spermatophore, visible at the base of the female's ovipositor.

This male **Empis tessellata** dance fly in England has presented his mate with a substantial nuptial gift, a large cranefly (Tipulidae).

KISSING FLIES

Flies of several families 'kiss', the female taking nutritious liquid direct from the male's proboscis during a usually prolonged affair while the male is mounted on the female's back. Several platystomatid flies do it, including the common black European *Platystoma seminationis,* as well as at least one species of stilt-legged fly (Micropezidae) and several tephritids, including *Paracantha gentilis* from California. However, several tephritids, such as *Eutreta sparsa* which lives on goldenrod plants in the USA, produce quite substantial meals of froth from the proboscis. This is not passed directly to the female in a 'kiss', but applied to a leaf in a blob or pillar, being gradually built up by the application of more and more material. Apparently the female needs to be satisfied that the male's frothy nuptial offering is large enough before she will consent to mate with the donor.

male issues an invitation to feed by raising his wings into a V position. The female steps forwards to sup directly from the male's proboscis in a 'kiss' which, though brief, is yet long enough for her to draw off the proffered blob of liquid. This is derived directly from the contents of the male's crop, and may yield considerable benefits in terms of larger eggs, whose production time is also reduced.

Male *Panorpa* spp., scorpionflies (Mecoptera), also deposit on to a leaf a blob of froth on which the female feeds as they sit in copula. In the related hangingflies (Bittacidae) the male presents his spouse with the corpse of an insect, guaranteed to be both tasty and nutritious as it has been subject to a quality test by its captor before being served up. Such nuptial gifts are not absolutely essential for the females, which are quite capable of catching their own prey. However, by providing a large and mouth-watering gift, the male can keep his mate happy long enough for full sperm transfer. During the height of the mating season the females no longer bother with the effort of catching their own prey, but rely solely on the free handouts from the males.

Female danceflies (Empididae) have no choice but to rely on nuptial offerings as a source of protein, for they do not themselves prey on other insects. Killing is the sole preserve of the males, who use their booty to ensure long mating times. Dance-flies earn their soubriquet from their habit of performing an aerial dance, during which the male hands over his nuptial gift and couples with the female. The female sucks the juices of the gift as they mate, turning it over and over and probing it repeatedly with her long proboscis. Most species dance in swarms, often around a marker such as a tree, which may be used year after year by successive generations. The large swarms of the small black European species *Hilara maura* frequently resemble swirling smoke and take place low above streams or ponds. The front legs of the male are distinctive for the grossly inflated metatarsi (situated just above the 'feet'), whose function is to produce silk for gift-wrapping a tiny insect. Like any system, this one is open to abuse, and males of some species present the packaging minus the gift.

05 Egg-laying and Parental Care

EGG-LAYING

Egg-laying methods in insects show a great variety. The methods available can be divided into the following:

- LIVE BIRTHS
- DIRECT DEPOSITION
  - by dive bombing
  - by ground-laying
  - by laying on plants
  - by laying on water
  - by countering hazardous defences
- INDIRECT DEPOSITION
- PAEDOGENESIS

Some of the more unusual egg-laying methods are outlined in the box.

SOME UNUSUAL METHODS OF EGG-LAYING

SPECIES	HABITS
Dythemis cannacrioides (dragonfly)	While perched, the female produces a large ball of eggs resembling a bunch of grapes. She then attaches these, in flight, to the roots of lianas where they hang down into tropical streams.
Prestwichia aquatica (chalcidoid wasp)	The tiny 1mm-long female uses its legs to swim underwater and parasitize the eggs of water beetles and aquatic bugs such as *Notonecta* backswimmers (water boatman).
Dermatobia hominis, human botfly (Calliphoridae)	The female attaches an egg to a biting fly such as a mosquito which carries it to the host. The host's body heat causes the larva to burst out of the egg and bore its way into the flesh, where it causes a painful swelling.
Galerucella viburni, leaf beetle (Chrysomelidae)	The female gnaws nest holes in a twig and then spends several hours carefully constructing a lid consisting of dung, glandular secretions and wood-chips.
Atherix ibis, fly (Rhagionidae)	The females gather on leaves over water to lay their eggs in huge groups. As each female dies soon after making her contribution, a huge pear-shaped mass of dead flies mixed with eggs soon builds up. The larvae fall into the water when they hatch.
Cryptoses choloepi, Panamanian sloth moth (Pyralidae)	Female moths live in the fur of arboreal sloths, quickly leaving their host to lay their eggs in its dung when it descends to the ground to defecate.

LIVE BIRTHS

Not all insects lay eggs. Live birth or viviparity has already been mentioned in aphids, and also takes place in a number of other insect groups, including leaf beetles (Chrysomelidae), longhorn beetles (Cerambycidae), cockroaches and thrips. The large chequered *Sarcophaga* blowflies (Calliphoridae) are usually in such a hurry to find a suitable corpse on which to lay their eggs, that these actually hatch in the female's uterus, from which emerge live wriggling maggots. The cockroach, *Diploptera punctata*, is currently the only known viviparous member of its group, but is especially fascinating. The female has developed internal glands akin to the milk-producing apparatus of mammals. It is upon this 'insect milk' that the baby cockroaches are succoured. A more common habit in cockroaches is for the female to carry her eggs around enclosed in a tough pod or ootheca, attached to the tip of her body. Some mosquitoes also carry their eggs around, attached to their legs, until a suitable egg-laying site is located.

DEPOSITING EGGS ON A HOST OR SUBSTRATE

DIRECT DEPOSITION

Normally, however, female insects introduce their eggs directly into a suitable environment via the ovipositor, which may be heavily modified according to the job it will be called upon to perform. Thus, in sawflies which lay their eggs in leaves or stems, the ovipositor has serrated edges for cutting a slit into the plant tissue. In tephritid flies the ovipositor is telescopic, and can be extruded to several times its original length, in order to probe deep inside the stems or flowerheads of plants such as thistles and knapweeds. Clearly visible ovipositors of often spectacular length and formidable appearance are found in many katydids, wood wasps and ichneumon wasps. The sabre-like adornment protruding at the rear of many female katydids looks capable of inflicting painful retribution to an enemy. In fact its sole function is to introduce the eggs into concealed places by probing into slits in bark, rotten wood, lichen-covered earth, cracks in the ground or plant tissue, according to species. In wood wasps (Siricidae), such as the impressive black and yellow giant horntail, *Urocerus gigas*, the long needle-like ovipositor drills into timber. The eggs are accompanied by a special symbiotic fungus, contained in a pair of sacs at the base of the ovipositor. To the wood wasp's larva the developing fungus represents a consumer-friendly diet, being easier to digest than the wood of the tunnels, which is also eaten.

The hair-like ovipositor in many ichneumon wasps, such as *Rhyssa* and *Certonotus*, may be several times the body length. It looks far too fragile to perform its allotted task of penetrating deep into the solid wood, of trees containing host larvae. In their galleries deep inside the wood these would appear to be quite safe from detection by any exterior enemy. However, the senses of the searching ichneumons are acute. The female *Rhysella approximator*, while combing alder trunks for its particular host (the alder wood wasp, *Xiphydria camelus*), delicately strokes the bark with down-curved antennae, apparently seeking to detect the faint signs of chewing emanating from within. With remarkable precision her hair-thin ovipositor is unleashed into the wood, sinking to its full depth in only a few minutes. If the plump soft defenceless wood wasp larva within is large enough to act as a host (an assessment the *Rhysella* makes using sense organs at the tip of the ovipositor), she will lay an egg. This both begins and ends its journey as a normal egg, but between it is distorted into a long thin strip as it passes down the ovipositor.

In most insects the actual mechanisms of locating a host, whether animal or vegetable, are still but little understood. For the females of many common hoverflies, such as *Metasyrphus corollae* and *Episyrphus balteatus*, whose larvae plunder colonies of aphids, the initial attraction is the scent of honeydew. This waste product is more or less constantly streaming forth from the perpetually feeding groups of their hosts, which can be of virtually any aphid species. In her turn, the parthenogenetic cynipoid parasite wasp *Callaspidia defonscolombei* uses the same aromatic cues for locating the syrphid larvae on which to lay her own eggs. In a number of tiny egg parasites, the female wasps cut out the need for days of tiring searching for isolated egg batches by hitching a ride on the adult host. When it finally gets around to laying its eggs, they make sitting targets for the tiny hitch-hikers.

The degree of effort invested in locating a suitable host is often quite manifest in many butterflies. Even in a suburban garden, it is easy to see a female of one of the common species, such as a large white *Pieris brassicae*, flying restlessly from plant to plant, landing and drumming briefly with the feet before moving on to the next leaf. The initial cues which provoke the brief landing-test are visual, with the shape of the leaf being important. Highly dissected leaves are harder to detect than large solid ones. Volatile odours given off by the plants are also probably used as cues by many butterflies. An example is the carrot volatiles which attract egg-laying in the American black swallowtail, *Papilio polyxenes*. However, some scents may actually deter landing and investigation, usually those given off when a plant has been damaged. These indicate the likelihood that it is already under attack by caterpillars, making it a poor prospect as a host due to larval overcrowding. Strangely enough, some mosquitoes actually practise the reverse tactic, preferring to lay their eggs in water containers which already contain mosquito eggs or larvae – but only as long as the latter are not too old. *Trichoprosopon digitatum* females incline towards ovipositing in a receptacle which already contains females guarding their egg batches. By contrast, receptacles holding larvae over eight day old are shunned, with the prospect that such large larvae will turn their attentions towards cannibalizing their newly arrived congenors. Most mosquitoes perch on the water's surface as they produce their neat rafts of eggs, but those which nest in rot-hole cavities in trees flip their eggs through the narrow entrance, in a precision dive-bombing exercise.

DIVE-BOMBING is also seen in beeflies, such as *Bombylius major*, which have to target their eggs into the entrances of their host bees or wasps. However, some beeflies rub their rear-ends in sandy soil, needing merely to lay their eggs close to the host colonies. The larva which emerges is able to make its own way into the nests. Once safely inside, it waxes fat on the stores of pollen and honey, so diligently laid in by the bee for her larvae. As these would compete with the invader for ownership of the foodstore, they are added to the menu, being gradually consumed, in such a way that death does not occur prematurely, which would leave a useless decaying corpse. Other beeflies develop as internal parasites of caterpillars or attack the buried egg-pods of locusts.

Brushing the tip of her abdomen against the sand of a coastal dune system in southern England, this bee fly, **Villa modesta** (Bombyliidae) deposits her eggs close to colonies of her solitary bee hosts.

GROUND-LAYING Locusts themselves are only large grasshoppers, a group in which egg-laying mostly takes place in the ground, using the extremely flexible telescopic nature of the abdomen. Indeed, this can be extended downwards to several times its original length. A few species, which have ovipositors capable of piercing plant tissue, lay their eggs inside stems. *Leptisma marginicollis* and *Stenacris vitreipennis* from the USA utilize giant bulrush stems. As they are produced, grasshopper eggs are covered in a layer of foam, which hardens as it dries into a tough envelope. Such protection is essential for the oothecae of most praying mantids, as these are usually placed in very exposed positions, such as branches, tree trunks or telephone poles. Many stick insects (Phasmidae) merely scatter-drop their large seed-like eggs in a very casual manner as they feed. However, some species take the trouble to flick them out, so that they land well away from the base of the food plant. This keeps them separate from faecal pellets, whose aroma is a powerful draw for certain wasps which parasitize phasmid eggs.

LAYING ON PLANTS Plants come to the unwilling service of insects in a rather complex way in the leaf-rolling weevils. In the hazel leaf-roller, *Apoderus*

This female sawfly, **Tenthredo scrophulariae**, is laying eggs in the leaf of a figwort plant **Scrophularia nodosa** in an English woodland. In sawflies, the ovipositor possesses twin blades which move back and forth with an alternating action, cutting a slit into which the egg will be inserted. This species is a Müllerian (unpalatable) mimic of **Vespula** spp. social wasps. These also happen to be the main pollinators of figworts, the only food plants for the sawfly. Both wasp and sawfly therefore gain mutual benefit through presenting a common defensive front around a point of shared interest.

A male and female great green bush cricket, **Tettigonia viridissima**, basking on low vegetation in south-west England. The long sword-like ovipositor of the female (bottom) is very obvious.

Bending her abdomen downwards, this female **Metasyrphus corollae** hoverfly (Syrphidae) deposits an egg amidst a thriving colony of the lupine aphid in an English garden *(below)*.

coryli, the female usually labours alone, although sometimes her mate will be perched on her back, ready to forestall any attempt by a rival to push in and mate at the vital last moment. After meticulously rolling up the leaf in a prolonged and pre ordained series of actions, the female lays one or more eggs in slits which she bites with her rostrum. In many weevils the rostrum is conspicuously elongated, bearing at its tip biting mandibles, capable of gnawing holes in various plant tissues, including wood. Tree trunks are the normal hosts of many longhorn beetles, in which the females use their powerful jaws to tear a groove in the bark prior to introducing an egg. The North American bark-girdler *Oncideres pustulatus* actually prepares her own supply of newly dead wood, by biting a ring around a branch, thus girdling it. She then bites out a series of grooves in which to insert her eggs.

WATER-LAYING Insects with aquatic larval stages often lay their eggs above or beside the water rather than in it, thus avoiding the considerable hazards posed by such voracious aquatic predators as dragonfly larvae, *Notonecta* spp., the water boatman and diving beetles. Many horseflies lay large clusters of eggs on vegetation overhanging water, into which the larvae drop as they hatch. Similar habits occur in many caddis-flies, in which the eggs are enclosed in a gelatinous mucilage which prevents desiccation in the exposed site.

Alderflies, such as the common European *Sialis lutaria,* also cement large egg masses to waterside plants. The first mass to be laid often seems to act as a powerful attractant to subsequent females, so that considerable aggregations are formed, covering the plants in a cladding of eggs. This egg-clumping may be a device designed to reduce the searching-effectiveness of tiny trichogrammatid wasp egg parasites. These are known to lay down chemical markers which enable them to recognize and avoid eggs which have already been parasitized, by themselves or other conspecific females. However, it takes time and a certain amount of effort to check this. Thus, when host eggs are placed in huge groups, the tiny parasites are bound to spend time rechecking used eggs, instead of locating fresh ones. The parasitic wasp, *Pimpla nipponica,* goes a step further by even recognizing hosts it has probed experimentally and then rejected, greatly reducing the time needed for searching.

With her mate still in tandem, an azure damselfly, **Coenagrion puella**, inserts her eggs into the stem of an aquatic plant in an English pond.

Some dragonflies also place their eggs above the water, but not in the large conspicuous masses seen in alderflies. The green and brown mottled pattern of the female southern hawker, *Aeshna cyanea,* blends in well with the rotting stump or gnarled waterside root in which she prefers to insert her eggs, one at a time through a short curved ovipositor. She usually chooses a spot that will become submerged when water levels rise in winter and spring. The eggs remain dormant, protected inside the wood until the following spring, when they hatch to produce aquatic larvae.

INSECT EGGS

Insect eggs come in a huge variety of shapes and sizes. Many are placed in the ground, in cracks in tree-bark or inside plant tissues where they cannot be seen. Others are placed in the ground, where they must be capable of surviving various hazards, such as torrential rain, prolonged drought or baking sun, and the ever-present risk from marauding insects such as ants.

The tough ootheca of a praying mantis, which is exposed in the open and highly resistant to weather and enemies.

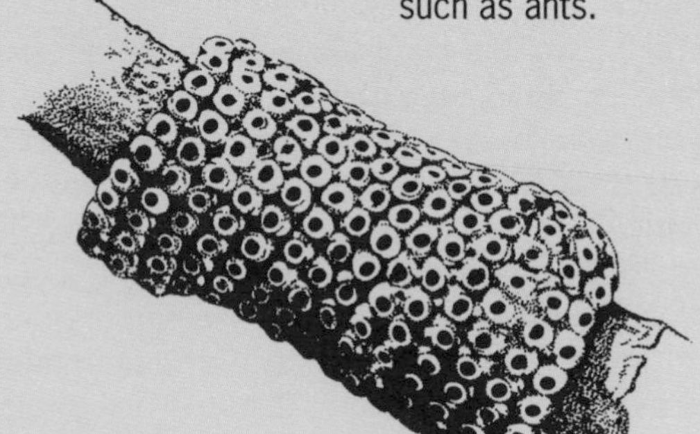

Lackey moth eggs ***Malacosoma neustria*** *placed around a twig like a collar.*

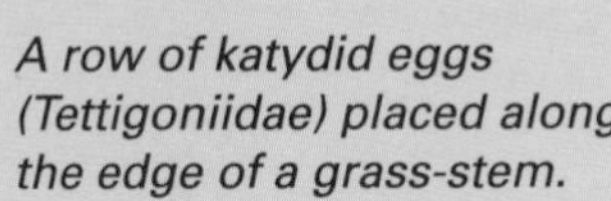

A row of katydid eggs (Tettigoniidae) placed along the edge of a grass-stem.

The large egg-mass made by a ***Tabanus*** *species horsefly on a sedge stem overhanging a pond. When the larvae hatch, they fall into the water.*

The bright yellow shiny eggs of the pot-bellied emerald **Gastrophysa viridula** leaf beetle (Chrysomelidae) are laid on the undersides of the leaves of docks (**Rumex** spp.). As in most leaf beetles the newly emerged larvae eat their egg shells before starting work on the leaf, which is soon reduced to shreds.

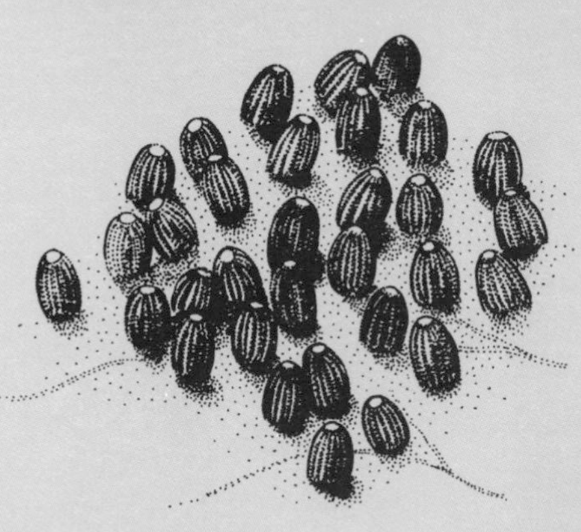

The ribbed eggs of the large white butterfly **Pieris brassicae** (left).

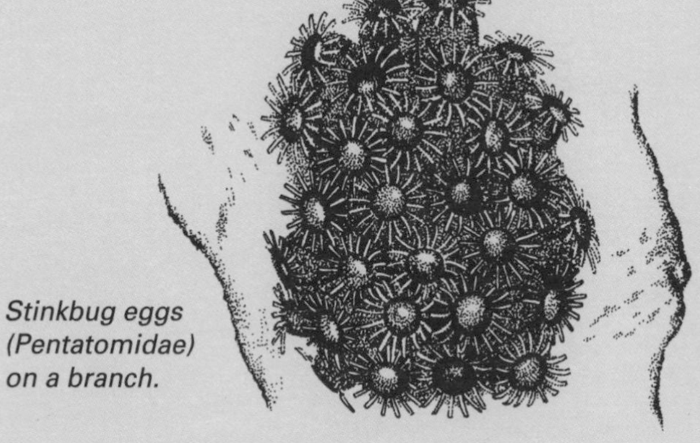

Stinkbug eggs (Pentatomidae) on a branch.

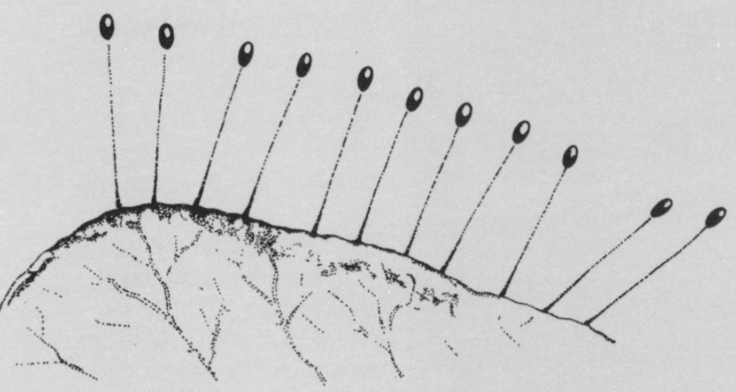

The eggs of green lacewings are placed on stalks. The female forms the stalk by oozing a drop of fluid on to a leaf and then drawing it upwards into a thread by raising her abdomen. The thread hardens almost as soon as it is formed, so that it can be topped off with the egg.

Many of the smaller darter and chaser dragonflies lay their eggs directly into the water. In the four-spotted chaser, *Libellula quadrimaculata*, the female dips rapidly up and down just above the water's surface, washing the eggs off the tip of her abdomen. Meanwhile, her mate hovers nearby, ready to mount a fierce resistance against any rival who tries to barge in and grab the female before she has laid her eggs. A more secure option is normal in the small darters, such as the common darter, *Sympetrum striolatum*. The male remains with his mate, attached to her in tandem by his terminal claspers, helping to lift her up and down as she strokes her abdomen against the water and releases her eggs. Male-assisted egg laying is also common in damselflies, which always insert their eggs into plant tissues, often partly or even totally submerging in order to do so. When the male remains in tandem, he not only guards against intruding rivals, but can also help to lift a chilled and waterlogged female clear of the surface, until she can warm up and dry out.

Calopteryx splendens females work alone as they submerge completely to lay their eggs in the stems of aquatic plants. After prolonged labours beneath the cool waters, they often struggle to take off from the surface, and become vulnerable to attacks by fish or other aquatic predators.

HAZARDOUS DEFENCES The females of many parasitic insects have to face a rather different challenge – how to get the egg or larva on to the host without

Willow sawfly larvae **Pteronus salicis** are capable of repelling an enemy by thrashing around and regurgitating a gob of sticky and unpleasant liquid. It therefore pays this ichneumon wasp to make her approach with extreme care. By standing at right-angles to her target she can stealthily bend her slender abdomen over to insert an egg, thereby reducing the chances of inducing a defensive reaction.

provoking a potentially hazardous defensive reaction. Even apparently harmless-looking hosts, such as caterpillars, can be more of a problem than could be expected. The usual reaction when a caterpillar senses a parasite actively showing an interest close by is to thrash wildly around. This makes it difficult for the parasite to find a foot-hold, and it may even be knocked bodily off for some distance. Many caterpillars can also bend their heads back at quite an acute angle, to daub a parasite with a nasty gob of regurgitated fluid. This can have fatal results for a tiny and delicate wasp. The parasite therefore has to act with extreme caution, pussy-footing around the hosts, withdrawing instantly if threatened and generally taking plenty of time. Many of the relatively large bristly-bodied parasite flies (Tachinidae) practise a more indirect and therefore less hazardous option, with the concomitant disadvantage of being less predictable. They carry out a rather hit-or-miss attack by widely distributing large numbers of very tiny eggs on the host's food plant. These hatch in the host's intestines after being unwittingly eaten. In other species, the eggs hatch soon after laying. The resulting larvae then bore their way into any suitable creature which comes within range.

Female *Lysiphlebus* spp. wasps, which parasitize aphids take advantage of the telescopic qualities of their abdomen. This is threaded forwards through stilted legs and extended fully until its tip touches an aphid. Such a cautious *modus operandi* might seem quite unnecessary against such an apparently defenceless target. However, aphids deploy their own version of superglue against their enemies. They secrete a waxy substance from the little spikes (cornicles) situated on the rear topside of the abdomen. Upon contact with something such as a parasite, the wax hardens rapidly, snaring the intruder or at least gumming it up sufficiently to cause it major problems. The most likely victims to make fatal encounters with the wax are tiny wasps such as *Alloxysta*. These have to brave the danger zone right up on the aphid's back, in order to probe with the ovipositor and locate the *Lysiphlebus* larva developing within. It is this, and not the aphid, which will act as the host for the *Alloxysta* larva. Such hyperparasites (a parasite of a parasite) are common in the insect world, and there are even hyper-hyperparasites!

Some parasites which attack hosts possessing formidable powers of reprisal seem to run extraordinary risks. For example, female conopid flies loiter near clumps of flowers which will be visited by bees, and then make a quick pounce in order to fix an egg to the bee's back. *Vespula* wasps receive the same treatment, despite the apparently suicidal nature of such a close approach to a creature blessed not only with a lethal sting, but also powerful meat-cleaving jaws. However, the female conopids are super efficient at this risky task, and seem to get away with it most of the time. As it develops inside a bumble bee, the conopid larva increases its chances of eventually becoming an adult fly by manipulating the behaviour of its host. Just before it is ready to pupate, the parasite induces the worker bee to dig down into the soil. This becomes a sanctuary where the fly's pupa can pass the winter, protected from enemies and the weather. Such a neat arrangement is far better than trusting to survival in whichever exposed spot the bee eventually happens to drop dead.

Several species of coffinflies (Phoridae) from the tropics lay their eggs on powerfully built ants, which are eventually decapitated from within by the developing fly larva. In the phorid *Metopina pachycondylae*, it is the ant larva which is attacked, eventually bearing around its neck an unwelcome guest in the form of the phorid's larva. Rather than browsing on its host directly, the guest acts as a cuckoo, stealing the victuals delivered to the ant larva by the workers.

INDIRECT DEPOSITION

Another way of effecting entry for a larva into a dangerous host is to do so indirectly. Cyrtid flies lay huge batches of eggs attacked to twigs. Upon hatching, the tiny larvae wait for a spider to come near, then leap on board and bore into its body, which they then devour slowly from the inside. Similar tactics are typical of many female *Meloe* oil-beetles, whose bloated abdomens are an overt indication of the gargantuan egg batches (between

3,000 and 4,000) which will eventually be attached to a leaf or stem. The resulting larva is fully legged and quite active, and is called a triungulin. Its first task is to climb up onto a flower and lie in wait until a solitary bee of the right genus chances along. After grabbing the bee's furry coat, the stowaway is ferried back to the bee's nest. Once safely inside this sanctum, the triungulin moults into a legless grub, which devours both the bee's egg and the enclosed provisions. The risks inherent in such a remotely targeted scenario are increased by the fact that the triungulin larva is not very fussy about what it fastens on to – the wrong type of bee, such as a bumble bee, or even a hairy fly will do. The chances of actually reaching the right nest and prospering are fairly remote, hence the need for saturation coverage with the huge number of eggs.

In the black-and-yellow wasp-like *Nomada* spp. nomad bees, the female is disguised in a cloak of chemical anonymity in order to gain direct access to her host bee's nest without being repulsed. Male nomad bees are able to synthesize the nest odour of their particular host, and apply this to their mate's body during copulation, thus using chemical mimicry to obscure her nefarious identity. Cuckoo-like habits are found in many other bees, although not all species penetrate the host nest through stealth. *Psithyrus* females, which resemble bumble bees, bulldoze their way in with a frontal blitz, killing any workers which stand in their way, and eventually adding the queen to the tally of victims. The cuckoo is able to bludgeon her way in like this because she has a more powerful sting than the nest's defenders, and is more heavily armoured.

The Trojan Horse strategy is the normal method employed by females of the muscid fly, *Stomoxys ochrosoma*, to infiltrate their larvae into the temporary bivouacs made by African driver ants (*Dorylus* spp.). The female fly hovers a few centimetres above a hurrying column of ants on the move, carrying brood to a fresh bivouac. She releases a clutch of some twenty eggs into the column, where they are picked up by a worker who is not carrying any burden. They thus arrive safely at their destination, delivered in the formidable jaws of the host ant. Once inside the bivouac, the fly larvae spend their lives as harmless scavengers. To try and enter the bivouac directly would undoubtedly be suicidal for the female fly, but her devious back-door methods succeed by exploiting the ants' natural instinct to carry something while moving house.

PAEDOGENESIS

Some insects never get around to laying eggs at all, because the second generation begins to develop inside the living body of the first, while it is still a larva. This phenomenon, known as paedogenesis, is particularly found in certain tiny gall midges (Cecidomyidae). The saga begins when a female lays a few extremely large unfertilized eggs. Each resulting larva contains within itself parthenogenetic eggs, which hatch to produce more larvae. These gnaw an exit from their doomed parent and start the whole process all over again. After this has gone on for several generations, there is a reversion to more conventional behaviour, with the production of pupae resulting in both male and female midges.

PARENTAL CARE

The vast majority of insects abandon their eggs once they have laid them. The task of ensuring that this is done in a suitable place, which gives them the maximum chance of successful development, is the end of the affair as far as the females are concerned. However, in a relatively small number of species the female's responsibilities extend beyond this to encompass guarding of the eggs and sometimes the young – or at least the preparation of a nest stocked with sufficient provisions for full larval development. In an even smaller band of insects, notably all known species of termites and ants, plus a relatively small proportion of the total known number of bees and wasps, care of the young extends to overlapping generations. This means that the developing brood is cared for by an

EXAMPLES OF PARENTAL CARE IN INSECTS

ORDER	SPECIES	HABITS
Blattodea	Brown-hooded cockroach (*Cryptocercus punctulatus*)	Forms family groups in rotten wood. The nymphs feed on predigested food in the form of gut contents exuding from the adult's anus.
Orthoptera	*Anurogryllus arboreus* (= *A. muticus*) (cricket)	The female prepares a brood chamber in the ground and stocks it with berries and grass. She lays special mini-eggs for use as baby food, and carries faecal pellets to a special lavatory chamber.
Embioptera	*Oligotoma ceylonica* (web-spinner)	The mother camouflages her eggs within silken brood-tunnels and is later followed around by her offspring on feeding forays.
Thysanoptera	*Anactinothrips gustaviae*	Forms dense colonies on tree trunks on which adults and nymphs form foraging bands. Females individually add their own contribution to communal egg dumps, which are guarded by several females.
Hemiptera	*Zelus* spp. (assassin-bug) (Reduviidae)	The male stands guard over an egg batch until the nymphs hatch. He then catches insects and proffers them to the young kebab-style skewered at the tip of his rostrum. Parental care by male insects is scarce, mainly because it can only pay dividends if the male has some fail-safe mechanism guaranteeing paternity of the eggs he is guarding.
	Platycotis vittata (tree-hopper)	The female guards her eggs and bites feeding-slots in the plant stem for the young larvae, so that their tiny mouthparts can obtain a meal.
Coleoptera	*Carterus caledonius* (ground-beetle)	The female excavates a labyrinth of cells below ground, each receiving its own individual egg. She stands guard over the sole entrance and furnishes her larvae with seeds from a larder which she has previously laid in.
	Spercheus emarginatus (water-beetle)	The sixty or so eggs are contained in a silken sac which the female attaches to her hindlegs. She carries them around like this, letting the sac hang down in the water to permit the maximum flow of oxygen, but swinging it up close to her body when she is threatened.
Lepidoptera	*Hypolimnas anomala* (butterfly)	Most (but not all) females stand on guard over their eggs and remain with their young larvae for the first few days.
Hymenoptera	*Themos olfersii* (sawfly) (Symphyta)	The female straddles her eggs and lunges at intruders. She then maintains a watch nearby as her larvae move around the food plant to feed.
	Cedria paradoxa (braconid parasite wasp)	The female keeps a vigil beside her offspring as they develop within the caterpillars of several pyralid moths.

PARENTAL BEHAVIOUR IN THE PARENT BUG

The parent bug, *Elasmucha grisea,* is a member of the large family of shieldbugs or stinkbugs (Pentatomidae) which exhibit many examples of parental care. This little bug is common on birch trees in Northern Europe, even colonizing solitary trees in strongly urban situations. The female lays a batch of some fifty eggs on the underside of a birch leaf. Although capable of producing a larger clutch, each female only lays the number of eggs which she can adequately protect, using her body as a shield. She is no mere passive guardian, but tilts her body towards an intruder and noisily buzzes her wings. Just how effective this is in repelling incursions by predators such as ants and bugs can be seen by removing the mother – all the eggs disappear within a few days.

When the nymphs hatch, their mother remains over them for a few days. She then stays close by as they move around to feed on the developing green seeds of the host tree. Mother and brood stay in touch via a trail pheromone which the nymphs lay down as they move. They can also release an alarm pheromone which brings their mother hastening to their aid. She finally dies in harness, just before her offspring moult into their last nymphal instar. Her role as a protector is

PARENTAL BEHAVIOUR IN THE PARENT BUG

redundant anyway, for the nymphs are finally big enough to take over their own defence. By sitting in a close-knit group, they can let fly with a salvo of chemical vapour of sufficient quantity and potency to send an enemy as large as a human reeling, sneezing and coughing.

After moulting into adults, the brood stays closely packed together on their birch leaf for a few more days, then disperse onto a wide variety of plants, before searching for a hideaway in which to pass the winter in hibernation.

earlier generation, some of which may be modified solely for this or other tasks within a highly complex social organization.

Simple care of the brood by the mother or father acting alone (for example, in bugs) or in consort (for example, in burying beetles or passalid beetles) is known as *subsocial behaviour*. Some examples of parental care are shown in the table on page 93.

PARENTAL DEFENCE MECHANISMS

In species such as earwigs and mole-crickets, which nest in damp subterranean chambers, a prime function of the mother is to lick the eggs regularly in order to stave off infection by moulds. This action probably applies a fungicide, although it is not essential in all species, for example some earwigs. The common *Forficula auricularia* is one of a number of earwigs exhibiting complex brood care, involving not just egg-licking, but even feeding of the young via regurgitation, as in many birds. The nymphs seem to beg for food by nuzzling the female's mouthparts with their own.

Egg-guarding is known in quite a number of treehoppers (Membracidae) and their relatives. Female thorn-bugs, *Umbonia crassicornis*, spend up to 30 days perched atop their fat clusters of eggs. The tiny newly-hatched nymphs are unable to penetrate the tough twigs of the food plant with their tiny mouthparts, so their mother prepares the way by cutting a series of slits in the bark. She is valiant in defence of her brood, and quite able to see off opponents such as ladybirds and even assassin-bugs, fanning her wings fiercely with an audible buzz. She may respond to an intrusion by detecting an alarm pheromone released by an injured nymph.

Such self-defence is not always necessary, however, as some membracids secrete enough honeydew to attract a mercenary defence force in the form of ants. If present in sufficient numbers, these vigilant and aggressive security guards can be so efficient that females of the North American membracid, *Publilia reticulata*, actually desert their eggs at an early stage, trusting them to the sole stewardship of the ants. By securing a premature release from her initial parental duties, the female has time to produce an extra 'bonus' clutch of eggs.

COLLECTING DUNG AS A LARVAL FOOD SOURCE

Mammal dung might be a throwaway substance as far as its former owners are concerned, but it is a benevolent windfall for the many insects able to make a living from the considerable body of undigested nutrients which remain. Unfortunately, dung has a habit of appearing sporadically and unpredictably, so insects which have a further use for it have acute senses, able to track down a fresh supply with considerable speed and accuracy. Haste is essential, because there will be a rush of competitors eager to cash in on such a bonanza resource. Each new arrival will have its own way of dealing with the competitive situation, the most interesting being the many kinds of dung beetles (Scarabaeidae). Their methods of handling the dung vary, but all are geared towards stashing away as much of the stuff as possible, as quickly as possible, before the opposition gets to work.

One approach is to construct a series of branching subterranean chambers directly beneath the dung. This then lies conveniently to hand to be pulled down below and moulded into a spherical, pear-shaped or sausage-like brood mass. Both sexes of the common European *Geotrupes stercorarius* cooperate to build a series of side galleries off a main tunnel, stuffing each one with dung to a depth of about 10cm to act as a larval foodstore. However, as several beetles, belonging to more than one species, may be digging away beneath a single cow-pat, overcrowding can lead to one larva accidentally breaking into another's burrow and wreaking havoc.

The dung-rolling scarabs solve the overcrowding problem by removing neatly formed balls of dung from the site of activity and burying them somewhere quieter. There are two kinds of ball: feeding balls for the adults' own delectation and brood balls for larval food. Male and female often team up to cut and mould the dung, but it is usually the male who pushes the finished product away, deftly scurrying backwards, propelling the ball with his back legs, the female perched on top. Not all species utilize dung. The North American *Deltochilum gibbosum* makes good use of other waste material, such as feathers and hair, which are moulded into a brood ball topped with a cover of rotten leaves and earth. Some South American *Canthon* spp. are able to make brood balls of both dung and carrion, while male and female *Cephalodesmius armiger* from Australia collaborate to produce a kind of artificial manure from leaves and fruits, mixed together with their own droppings. The female constantly checks on the larva's food supply, adding fresh material as and when required from a central foodstore.

Having established a substantial cache of rich food in a relatively safe subterranean burrow, each female dung-beetle can afford to reduce the number of offspring she produces to a very low number – often only two to four – which she will have time to nurture most carefully. Compare this with the 3,000 to 4,000 eggs deemed essential to beat the odds in the oil-beetle 'spot-the-right-bee' lottery. A reduction in the number of offspring, whose future is relatively assured, is typical of most insects, such as solitary bees and wasps, which invest considerable effort in nest preparation.

BURYING BEETLES

Within the Coleoptera, the pinnacle of parental solicitude is arguably found in the *Necrophorus* burying beetles. A small animal corpse, such as a mouse, is the trigger for a complex suite of behaviours, usually performed by a male and female working together. With time not on their side, the pair toil away to scrape and bite the fur off the corpse as they gradually inter it inside a 'crypt' beneath the ground. Once safely hidden from competitors, the nice clean-shaven meaty corpse can now easily be moulded into a more or less globular shape, in which the female chews a bowl-shaped depression. It is into this that the larvae eventually crawl, sticking out their tiny heads and begging for food from their parent's mouths, like chicks in a nest. Parents and offspring also communicate with one another through a series of chirps. In the North American *Necrophorus orbicollis*, which normally nests in pairs, a single parent can respond to the loss of its mate by increasing the amount of time spent feeding the larvae. Such sophisticated behavioural flexibility has so far not been observed in any other invertebrate species.

BARK-BEETLES

Bark-beetles in the families Passalidae and Scolytidae also make exceptionally caring parents. These tiny insects inhabit galleries in timber, which they deliberately infect with a special species of fungus. This both predigests the wood into an edible state and actually acts as food for the beetles. Fungus-cultivating scolytids even have specific pouches for ferrying a supply of fungus to new trees. The male of the pine engraver beetle, *Ips pini*, spends several weeks in the breeding galleries, keeping them clean by removing his mate's droppings and helping to expel parasites. His presence makes a vital contribution, as fewer young are reared if he is removed. Some passalid beetles have evolved remarkably complex social interactions, with more than one generation living together as an extended family in the breeding galleries. A young adult will even pull its weight by helping to construct the complex pupal cases needed for a new generation of brothers and sisters. In *Monarthrum*, the female responds to the larva's needs by placing fresh supplies of fungal food at the mouth of its burrow, removing its droppings on a regular basis.

MASS PROVISIONING AND PROGRESSIVE PROVISIONING

Few female solitary wasps actually overlap with their offspring in a way which enables a decision to be made on whether or not fresh provisions are required, but this does happen in *Bembix* and some *Ammophila*. Usually female bees and wasps restrict their duties to building a secure nest and laying in sufficient provisions to last the larvae until they pupate. Long-term food-storing of this type is called mass provisioning and is typical of the spider-hunting wasps (Pompilidae). Sometimes the wasp's sting merely induces a temporary paralysis in the spider, allowing a safe margin to lay an egg on its body and depart. The wasp's larva then gradually consumes the spider's substance from without, eventually killing it. However, most pompilids sting their prey with sufficient severity to procure permanent paralysis, enabling the still-living spider to function as fresh meat for the larva in a burrow dug by the female. Similar habits are found in most sphecid wasps, with the prey varying from grasshoppers to caterpillars and even bees. Most species prey on only a single type of insect, such that *Cerceris arenaria* from Europe will only take weevils, although any type of weevil will do.

Bembix and *Stictia* confine their attentions to flies, and once again, any kind will do, from agile slender-bodied robberflies (Asilidae), fierce predators in their own right, to plump and rather clumsy flesh-flies (Sarcophagidae). These are supplied to the larvae on an as-needed basis called progressive provisioning. The female knows when more food is required because her first chore each morning is to visit the burrow to carry out a catering check and cleaning exercise. Finding the larder empty does not necessarily inspire a hunting foray. Some females stay behind near the massed burrows and adopt a body-snatching routine, pouncing on females returning from a successful, but time-consuming, hunt and trying to steal their hard-won fly. Not surprisingly, the owner reacts with some vigour, so that savage fights frequently break out near the nests.

PARENTAL CARE IN SOLITARY BEES

Solitary bees greatly outnumber the social kinds, and use a fascinatingly wide range of nesting mate-

A **Megachile** sp. leafcutter bee takes a section of leaf into her nest in an old wall.

A dune snail-bee, **Osmia aurulenta**, devotes many hours of meticulous labour to closing her completed nest with a mastic curtain made from chewed-up leaves. This species is common all round the coasts of the British Isles.

rials. Leaf-cutter bees use their strong jaws as shears to snip away neatly shaped pieces of leaves which will form a cigar-shaped cell. The nests are usually placed in a cavity such as an old beetle boring, but holes in walls or old nail-holes in woodwork are also used. Leaf-cutters are therefore often common in urban situations, and can even become a nuisance when they insist on excavating nest holes in the compost of glasshouse pot-plants. The females cut oval leaf sections for the sides of the cell, and circular ones for the top, 'knowing' at which point to switch from one to the other.

Mason bees employ more wear-resistant material for their cells. Many species use mud mixed with saliva to build highly durable exposed structures, quite capable of weathering the worst that winter storms can throw at them. Others, such as the red mason bee, *Osmia rufa,* nest in ready-made cavities, partitioned into cells using mud gathered from miniature 'mines' and ferried back in the jaws. Making further use of a prefabricated home, vacated by its former owner, still in good condition, but in need of some renovation, is the sensible policy adopted by several bees which nest in empty snail-shells. On coastal sand-dunes around the British Isles such convenient homes often lie around in their thousands, offering plenty of scope for hard-working *Osmia aurulenta* females to fill them with cells made from a mastic of chewed-up leaves.

Another plant material which is readily available is resin, which often oozes from trees such as pines. A wide variety of bees use this malleable substance as a building material within an existing cavity, either alone or mixed with other materials such as sand or woodchips. The wool-carder bee, *Anthidium manicatum,* collects the down from hairy-leaved plants such as mulleins (*Verbascum* spp.). The female scrapes up a considerable ball of fluff, bigger than herself, and uses it to line her nest cells. Like most solitary bees, the female provisions each cell with a mixture of pollen and nectar, and then abandons her handiwork, never actually seeing her offspring.

A step forward from this solitary lifestyle is to remain in the nest long enough to recruit newly-emerging daughters as helpers. This has happened in a number of species, such as the European bee, *Lasioglossum malachurum*. An overwintered female emerging in spring builds and provisions a series of nest cells from which emerge a bevy of daughters. These are rather smaller than their mother and, most importantly, are infertile and incapable of breeding. They therefore have a powerful reason for staying in the nest and devoting their lives to helping to raise more brothers and sisters. They take over the task of constructing and provisioning new cells, leaving their mother to concentrate purely on laying eggs. Towards autumn fertile males and females are produced, so that inseminated females will be available to continue the cycle next spring.

Bumble bees (*Bombus* spp.) have reached a similar stage of social development. An overwintered queen, awakened by the warm spring sunshine, devotes several weeks of hard labour to building and stocking a small number of nursery cells. She has to visit flowers for her supplies of nectar and pollen, but constructs the cells themselves from wax produced by her own body. The first adults produced from such inadequate means are always very small, due to larval undernourishment, and are all females. They nevertheless set about the task of helping to build the next set of cells, freeing the queen to concentrate on egg-laying. Unlike in *L. malachurum*, the queen's daughters are not sterile, and will eventually start laying eggs. Initially these will be eaten by the queen, but as the colony gets larger, the number of eggs laid overwhelms her ability to find them, and they then yield larvae. However, as the daughters were never inseminated by a mate, only male bees eventually emerge (due to the special hymenopteran system of producing females only from fertilized eggs). In late summer the males produced in this way leave the nest and mate with queens. These will have been specially reared, probably through being allocated extra-large rations of food by the workers.

SOCIAL INSECTS

The above bees, along with a number of others and a few wasps, such as *Parischnogaster nigricans serrei* from Malaysia, thus exhibit a form of social organization which entails sterile non-reproductives or 'workers' and a division of labour. A further step up the social ladder has produced the fully social bees, wasps, ants and (in a separate order) termites, in which several generations overlap in nests which can last many years and reach a huge size. The major characteristic of many highly social insects is the production of separate castes, designed for carrying out specific roles. These castes can be very obvious on account of their very different appearance, such as the huge head-capsules and mandibles of soldier ants, or the nozzle-shaped heads of some soldier termites. These are suited only to spraying defensive liquids at an intruding ant, and are quite unable to perform the precision building tasks carried out by the normal-jawed workers.

SOCIAL BEES

The most highly social bees are the seven species of honeybees in the genus *Apis*, plus the stingless bees. Colonies of the familiar hive bee, *Apis mellifera*, may contain up to 80,000 individuals, dominated by a queen who can mass produce as many as 2,000 eggs per day. As each worker bee ages, the tasks it performs on behalf of the colony gradually change in a predetermined pattern. A typical young worker will spend the first three-weeks of her six-week life as a domestic servant, feeding larvae and the queen, constructing new cells from her wax-producing glands, performing the housework necessary to keep the nest in a sanitary condition, capping larval cells and cleaning out old ones for reuse, and receiving nectar from incoming workers. She semi-ripens this by exposing it to the air in her jaws, before storing it in special cells. Any worker may get involved in fully ripening the honey by blowing air across it with rapidly vibrating wings. A similar behaviour is also brought into play for ventilating an overheating nest. If this fails and the temperature continues to rise, the workers sally forth to collect emergency supplies of water. Once spread over the surface of the comb, this will assist with evaporative cooling. Most other social wasps and bees employ similar crisis tactics.

PARENTAL CARE IN POTTER AND MUD-DAUBER WASPS

Potter and mud-dauber wasps are interesting for the resilient structures which they so skilfully fashion. The habit of constructing mud nests has evolved independently, potter wasps belonging to the Vespoidea (along with the familiar and notorious social wasps), while mud-daubers belong to the Sphecoidea, which contains most of the solitary nesters. Potter wasps often affix their flask-like nests to leaves or stones, usually filling them with partly paralysed caterpillars as food for the young. Mud-daubers tend to stock their nests with spiders.

continued overleaf

The finished pot is stocked with semi-paralysed caterpillars whose faint movements might damage an egg merely deposited on top of them. Therefore the female wasp inserts her abdomen and lays an egg on a stalk which keeps the egg and newly-hatched larva out of harm's way.

A potter wasp delicately moulding the lip of her pot with her mandibles.

Solitary wasps often steal one another's prey, a behaviour which is classified by biologists as kleptoparasitism. **Ammophila aberti** which nests on salt-flats in Utah is particularly notorious for this habit, epitomized by these two females disputing the ownership of a caterpillar which one of them was bringing back to her nest (*right*).

PARENTAL CARE IN POTTER AND MUD-DAUBER WASPS (continued)

Sceliphron spp mud-dauber wasps hold their ball of mud between their jaws and front legs as they fly back to their nests.

Among the less highly developed of the social wasps, the uncovered nests of *Polistes* and related genera are the easiest to study. This leaves the cells open so that the day-to-day workings of the nest are clearly visible. In genera such as *Parapolybia, Belonogaster, Mischocyttarus* and *Polistes* the original nest is founded by just a single queen, working alone for a time until a few cells have been constructed. She is then often joined by several more females, following which a phase of savage fighting establishes a dominance hierarchy which decides who will lay all (or at least most) of the eggs. The inner queen (usually the original nest-foundress) then continues to exert her influence over her 'workers' through constantly keeping them on edge. She does this through a regular regime of harassment involving frequent bitings and maulings, reminding her nest-mates who is boss. It thus falls to the subordinate females to carry out most of the daily routine on the nest. They forage for food, bringing in neatly rolled-up packages of caterpillars or other insects. These are then shared out with one or more of the other 'workers', who chew or 'malaxate' the food for a while, extracting some of the liquid contents for their own nourishment. The remaining bulk is proffered to the larvae, who stick their heads out of their cells when the adult announces feeding time by rapidly drumming on the nest with her abdomen. The provision of food in a processed rather than 'whole' form is typical of social wasps, and marks them out from the solitary wasps who stock their cells with complete insects. In addition, social wasps do not sting their prey in order to subdue it, as in solitary wasps, but inflict mortal damage with their powerful jaws.

Social wasps, such as this **Polistes buyssoni** in Chile, construct their carton nests from 'paper', a mixture of saliva and wood shaved off dead trees, fence-posts or woody stems.

In most of the highly social wasps the nest is made of so-called paper, a mixture of saliva combined with wood scraped off fences, gates or the dead stems of certain plants. In the highly social hornets and yellowjackets (*Vespa* spp. and *Vespula* spp.) the queen always founds a nest solo, as in bumble bees. Her subsequent dominance over her daughters is through chemical rather than physical means. The surface of her body is bathed in a pheromone which is eagerly licked off by the

PARENTAL CARE IN POTTER AND MUD-DAUBER WASPS (continued)

This **Polistes testaceicolor** (paper wasp) on its nest in the gloomy understorey of a Peruvian rainforest is malaxating a bolus of food which it has taken from an incoming forager.

Polistes wasp constructing a new cell from masticated wood-pulp or 'paper'. She gauges its size and shape by using her antennae as calipers to measure the reference points with neighbouring cells (*right*).

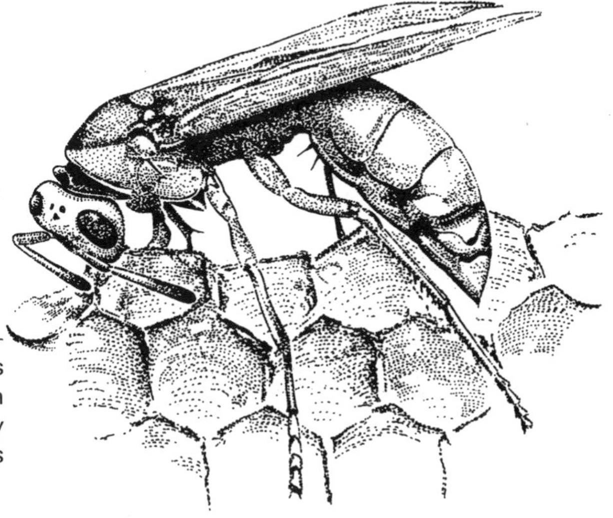

workers, thereby ensuring their continuing obedience. A similar royal pheromone also initiates the start of construction of specially large 'queen cells' by the workers. When the larvae are hungry they call for food by scraping their mandibles against the cell walls.

With her short life now half over, the worker is at last free to leave the nest as a forager. If she finds a patch of flowers rich in nectar and pollen, she can communicate its location and quality by a special 'dance' performed on the comb (*see* panel).

The queen maintains her dominant role through the production of a special 'queen substance'. A 'court' of workers pick this up by contact with her and spread it around to all the other workers. When the nest becomes too big for this vital chemical inhibitor to make the rounds on an essential 10-hour basis, then some of the workers start to rear new queens. The first task of a newly emerged queen is to sting to death all her royal sisters lying helpless in their cells. She then leaves the nest to mate and, shortly after her return, claims the nest as her own. This is because, surprisingly enough, it is the queen-mother who now forsakes her former home to found a new one, accompanied by a swarm of workers. In stingless bees, it is the young queen who establishes the new nest, leaving her mother, who has for a long time been incapable of flight, still in charge of the old one.

BEE DANCE

In *Apis mellifera* the dance is a stylized re-run of the foraging trip. With an audience of 'followers' (special dance-interpreters) crowded around her in the darkness of the hive, she begins her routine. For a food-source within 25m of the hive, she uses a 'round dance'. This involves a number of circular runs incorporating numerous switches of tack. The more switches, the richer the food.

To indicate food sources more than 100m away, the bee performs a 'waggle dance', covering a reduced figure of 8. During the top and bottom half-circles of the '8' she runs ahead, but along the central straight cross-pieces she breaks into a waggle. The longer the straight run and the more numerous the waggles, the greater the distance to the food. As the dance is performed on the vertical comb, she communicates direction by the degree of deviation from the vertical during the straight run. This indicates the angle between the food-source and the sun as perceived by a departing forager. The bee's inbuilt body-clock even compensates for the degree of movement observed by the sun during the period of her dance.

ANTS

Ants are unique within the Hymenoptera in that all known species are social – there are no solitary ants. However, some ant colonies are less highly developed than others, and some do not even have permanent homes. The army ants (*Eciton* spp.) of South America and driver ants (*Dorylus* spp.) of Africa lead a nomadic way of life. In a massed frontal attack they scour the forest for animal food, bringing it back in dense columns to a temporary bivouac housing the queen. With so many workers available, even quite large prey items such as giant centipedes can be manhandled back to the bivouac in one piece, rather than being sectioned up at point of capture. If a long-established wasps' nest comes under siege, the expected battle fails to materialize, as the formidably armed defenders merely hover nearby, impotent before the sheer weight of numbers of the raiders, who are left unmolested to rifle the cells of their precious contents of larvae and pupae.

A less destructive relationship with their fellow insects is practised by the ant *Dolichoderus cuspidatus* from Malaysia. Although also living in temporary bivouacs, these contain not only the ants themselves, along with their larvae and pupae, but also herds of the mealy bug, *Malaicoccus formicarii.* The ants regularly carry these out to pasture on fresh young plant growth near the bivouac. When the distance becomes too great to be economic, the ants break camp and move off *en masse* to a new site, carrying their domestic bugs with them. In common with the army ants, they form living bridges of workers across gaps along their path.

Sociality in ants is strongly based on the caste system, with predetermined bodily size and structure immutably dictating the lifetime role which any nest member is destined to play. Soldiers are often larger than workers, with heavier mandibles and the

With two minim workers 'riding shotgun' on its leaf-segment, an **Atta cephalotes** worker hurries back to the nest in a Costa Rican rainforest.

expanded muscles necessary to work them are contained inside an often grotesquely expanded head capsule. The specialised vocation reserved for the tiny minim workers of the leaf-cutting ant, *Atta cephalotes*, is to ride shotgun on the leaf fragments being carried by average-sized workers. The minim's duty is to repel parasitic phorid flies which try and use the leaf as a foothold in order to place an egg in the head-capsule of the carrier. The most bizarre members of the caste system are perhaps the repletes of the honeypot ants. These spend their entire lives as swollen storage containers for honey. They hang immobile from the roof of the nest, regurgitating some of the contents of their grossly distended abdomens on demand. The mouth-to-mouth passing of liquid food (trophallaxis) is a vital part of the social system within most ant colonies, and is also how the larvae receive their rations.

TERMITES

Termites are rather ant-like both in appearance and behaviour, with all species being social. However, the oft-used term 'white-ants' is completely incorrect, as termites are not related to ants, but belong to a separate order, the Isoptera. A major difference between the organization of ant and termite societies is the fact that workers in a termite nest are not exclusively female, but comprise equal numbers of both sexes. Unlike in the Hymenoptera, termite nests are a joint venture between a 'king' and a 'queen'. These embark upon their long reproductive careers by using secretions from within their own bodies to raise the first vital handful of workers. As the colony-size snowballs, the 'king' remains constantly by his mate's side, ready to top up her sperm supplies depleted by the continuous production of fertilized eggs, which may attain 30,000 per day. In social hymenopterans, by contrast, the queen mates on a single day, and the sperm from this single mating (or multiple matings in some, such as the hive bee) is stored live within her body and eked out over the years. It is sufficient to fertilize all the prodigious number of eggs that she will lay within her long lifetime.

Most termite workers are blind, and are divided into different castes specialized for roles such as foraging, nest repair, brood-raising and defence of the colony. The complex nests, with their sophisticated ventilation systems, are a testimony to the pinnacle of achievement available to those who adopt a social organization.

06 Food and Foraging

When considering diet, it is important to remember that in species which exhibit complete metamorphosis (for example, beetles, flies and butterflies) the food sources tapped by larva and adult are usually radically different. Thus, a spotted longhorn beetle, *Strangalia maculata,* licking nectar from the exposed nectaries of a hogweed (*Heracleum sphondylium*) flower in a woodland, will have spent its larval life chewing patiently away at a barely digestible diet of timber inside a tree.

Some insects have in fact passed their entire feeding activities over to the larval stage, forsaking food entirely as adults, with little choice in the matter, as the adult's mouthparts have degenerated and become inoperable. That impressive giant among flies, *Pantophthalmus tabaninus,* will therefore fast during its entire adult existence, having developed from a huge larva which dines on a rich, if smelly, diet of fermenting sap, deep within the spongy, almost liquifying wood of a fallen rainforest tree. When a trunk is richly colonized by these larvae, they announce their invisible presence by the constant slurping and squelching sounds issuing from within and audible from more than a metre away.

The aquatic larvae of *Toxorhynchites* spp. mosquitoes are voracious predators, gorging themselves on a nutritious supply of other mosquito larvae. The attractive metallic adult can therefore produce a full batch of eggs without the need to resort to a rather risky blood meal – a contrast to mosquitoes having vegetarian larvae. *Toxorhynchites* adults can sustain their activities quite adequately on a diet of nectar from flowers. This is rich in energy but deficient in the proteins needed for the costly production of eggs.

Mayflies (Ephemeroptera) take matters to extremes by passing a whole year underwater as vegetarian nymphs, only to allocate a few hectic hours to an adult life devoted to a frenzy of sexual activity, during which no food is taken due to atrophied mouthparts.

PLANT-PREDATORS

The vast majority of insects exploit plants in various ways. However, the methods by which this is achieved are varied and often complex.

- LEAF CHEWING
- SAP SUCKING
- SEED FEEDING
- ROOT CHEWING
- BARK AND WOOD BORING

Leaves are chewed by a host of insects belonging to many orders but especially by the larvae of moths and butterflies, by leaf-beetle (Chrysomelidae) adults and larvae and by grasshoppers and katydids. The tiny larvae of many flies, moths and beetles actually squeeze into the space between the upper and lower surfaces of a leaf, performing gastronomic tunnelling exercises to form mines of characteristic shape, according to species. Sap is sucked by many bugs, such as tree-hoppers, aphids and cicadas. Seeds succumb to inroads by the larvae of

ARRIVING WITH A BANG

Larvae of the European gorse weevil, *Apion ulicis*, destroy up to 80 per cent of the seeds of its host plant, *Ulex europaeus*, in a given area. The adult weevil is incapable of making its own exit from the ripe pod, so has to bide its time until its home self-destructs naturally, bursting with an audible crack which shoots the weevils out explosively, along with any surviving seeds.

HELP WITH DIGESTION

Most wood-boring insects, as well as many which eat leaves, are incapable of making fully productive use of their food without the help of cellulose-digesting symbiotic protozoa and yeasts. This is because few insects possess digestive enzymes capable of breaking down the cellulose content of wood and the cell walls in leaves. The essential digestive aids are often passed from mother to offspring as the eggs are laid, at which point they are smeared with a dose of the relevant symbiont. It is for this reason that the first meal of many beetle larvae is their eggshells. In other species, such as many weevils, the symbionts infiltrate the eggs while they are still in the female's body. Lepidopteran caterpillars, by contrast, cannot make use of cellulose and are obliged to prosper solely from the less nutritious cell contents. The prodigious appetite which results has unfortunate consequences for the plants, which may be denuded of their leaves over wide areas by species such as processionary moth larvae or American *Malacosoma californicum* (tent-caterpillars).

certain moths and by numerous beetles, especially weevils. Many bugs pierce and suck seeds, both fallen and while still in the pod. They inject a powerful digestive enzyme through the rostrum and into the seed, partly predigesting the kernel so that it can be sucked up in a convenient liquid form. Small nymphs of seed-suckers such as cottonstainers (*Dysdercus* spp.) often cluster around a single seed in a combined assault to infuse sufficient enzyme to liquify its contents. Roots comprise the dietary fare for a legion of subterranean chewers such as

The fruits and seeds of plants are often destroyed before they are even ripe. This green fig-eater beetle, **Cotinis mutabilis**, is eating an agave fruit in the Arizona Desert, USA. Scarab beetles take a wide variety of food, including fermenting tree-sap, leaves, flowers and dung.

Shortly after hatching in a Trinidadian rainforest, these tiny **Edessa** sp., stinkbug (or shieldbug) nymphs (Pentatomidae) will acquire a dose of essential symbiotic bacteria by feeding on their old egg-shells. **Edessa** spp. are generally warningly coloured in all their stages, feeding on poisonous host plants such as members of the potato family (Solanaceae).

cockchafer beetle grubs (*Melolontha melolontha*), wireworms (larval Elateridae) and cranefly larvae (Tipulidae). The latter are the well-known leatherjackets which feed in large numbers on the roots of grasses, including on lawns. Root-suckers include many aphids, and the larvae of cicadas, some of which may spend many years underground before their rather meagre diet has yielded enough nutriment to enable the transition into the winged adult. Finally, the tough bark and heartwood of trees is attacked by a whole range of beetles, especially longhorns (Cerambycidae), as well as by goat moth larvae (*Cossus cossus*) and horntail wasps, such as the giant horntail, *Urocerus gigas*.

In many instances, such as the pollinating services performed by insects, the plants gain something in return, rather than being just the losers. They have evolved ploys, such as physical and chemical defences, aimed at conferring the benefits of being winners all the time, only to lose any temporary advantages to some particularly adaptable insect attacker. Thus, the leaves of the 'mala mujer' (*Cnidoscolus urens*) from Mexico are armed with an array of stinging hairs which pack an agonising punch. When damaged, they also secrete a sticky latex sufficient to glue most attacking insects firmly in place. The caterpillar of the hawkmoth (*Erinnyis ello*) circumvents both these defences by grazing an area clear of stinging hairs, while removing any latex from its body with its mandibles. It then cuts off further supplies of latex by inflicting a thorough mangling to the leaf stalk with its jaws.

Most milkweeds (Asclepiadaceae) stockpile copious amounts of latex within ducts which run parallel to the leaf veins. This latex is not only toxic but also hardens rapidly upon contact with air, gumming up the mouthparts of an attacking insect. By severing the leaf veins, or more especially the

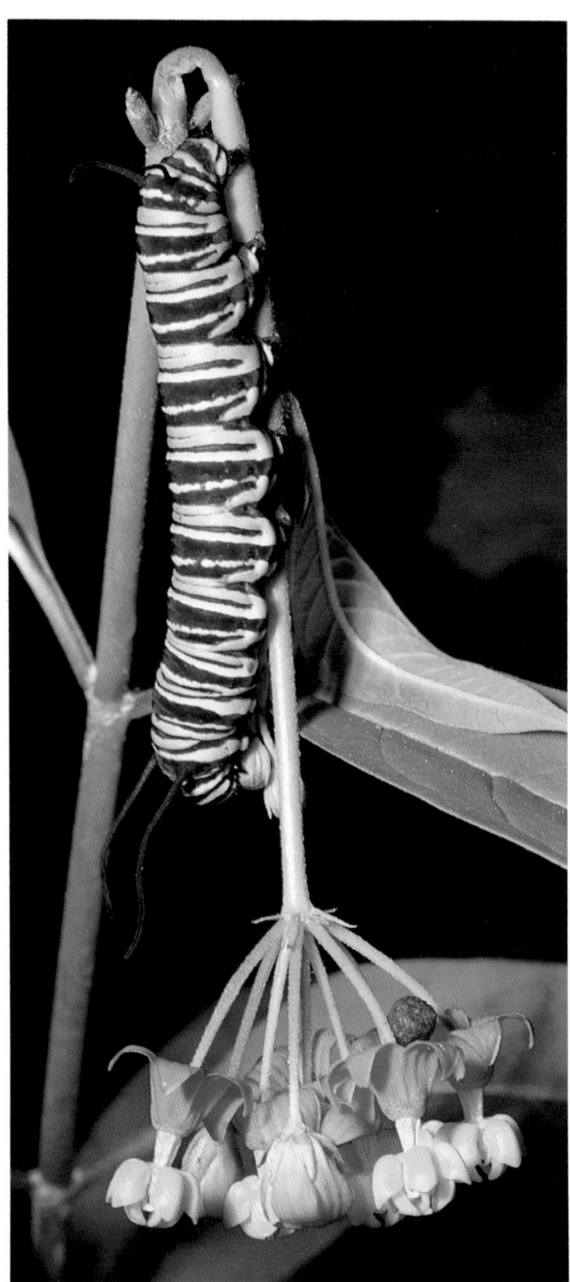

Redleg, **Asclepias curassavica**, is a particularly toxic member of the milkweed family, Asclepiadaceae. In order to cut off the supply of latex to the part of the plant on which it is feeding, this monarch butterfly larva, **Danaus plexippus**, has chewed a broad trough in the stem, causing it to sag.

mid-rib or even the leaf stalk itself, the supply can be cut-off, and this is always the first act performed by a number of insects which specialize in feeding on milkweeds. Small caterpillars of the Queen butterfly (*Danaus gilippus*) first chew out a circular furrow which acts as a moat, isolating a safe feeding area within.

Much as the human body sends white blood cells and antibodies speeding to its defence when invaded, some plants can respond to attack by diverting deterrent chemicals to the site where damage is occurring. Squash plants take about forty minutes to mobilize fully their defences but with little effect if the assailant is an *Epilachna* beetle, which first chews a trench to act as a buffer against the incoming flow of poisons. It really seems that the plants simply cannot win, even when they deploy highly toxic chemicals, such as cyanide, with its potent and usually fatal effects on animals. The lima bean (*Phaseolus lunatus*) contains relatively massive amounts of this dangerous substance (as linamarin), yet has only succeeded in increasing its attractiveness as food to one of its major enemies, the southern armyworm, *Spodoptera eridiana* (Lepidoptera: Noctuidae). Caterpillars of this remarkable moth actually need the presence of cyanide as a feeding-stimulant and thrive better on plants containing cyanide than those without.

About 10 per cent of aphid species have developed a specialized life-cycle based on alternation of their host plants. This involves spending the winter as an egg on a woody (primary) host, such as a tree or shrub. The egg hatches to produce a female known as a fundatrix. After feeding for a while, she migrates in early summer to a rapidly growing softer-leaved herbaceous plant or weed. This is an extremely hazardous undertaking, such that only perhaps one in ten thousand of these tiny pioneers will make it to their new host. Despite such unattractive odds, it seems to be worthwhile in view of the very advantageous nutritional rewards offered by the actively growing new (secondary) host. With ample provender to hand, the relatively few fundatrices, which win through to their destination, now generate a phenomenally rapid population boom via parthenogenetic reproduction. Few

GALL-FORMING INSECTS

Insects in several orders, but especially gall-wasps in the family Cynipidae, seem to exploit plants in an exceptionally high-tech way by actually 'instructing' them how to behave in a manner beneficial to the larvae. It is thought likely that the egg-laying female actually injects a substance which can interfere directly with the plant's DNA, thereby compelling the cells to multiply abnormally. This produces a swelling whose shape and colour is characteristic of the species concerned. The gall gives physical protection to the larva within, which finds itself in the position of being able to eat its home, without ever depleting its supplies, as the gall's growth keeps pace with its occupant.

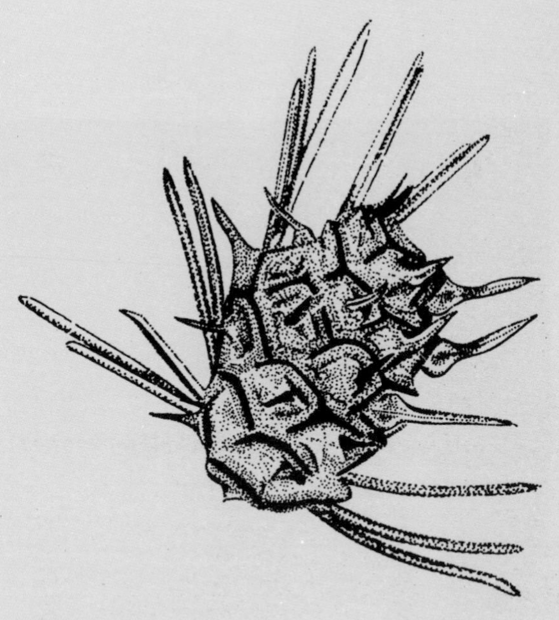

Pseudocone gall caused by the aphid **Adelges abietis** on spruce. The gall is caused by a basic part of the needles swelling up.

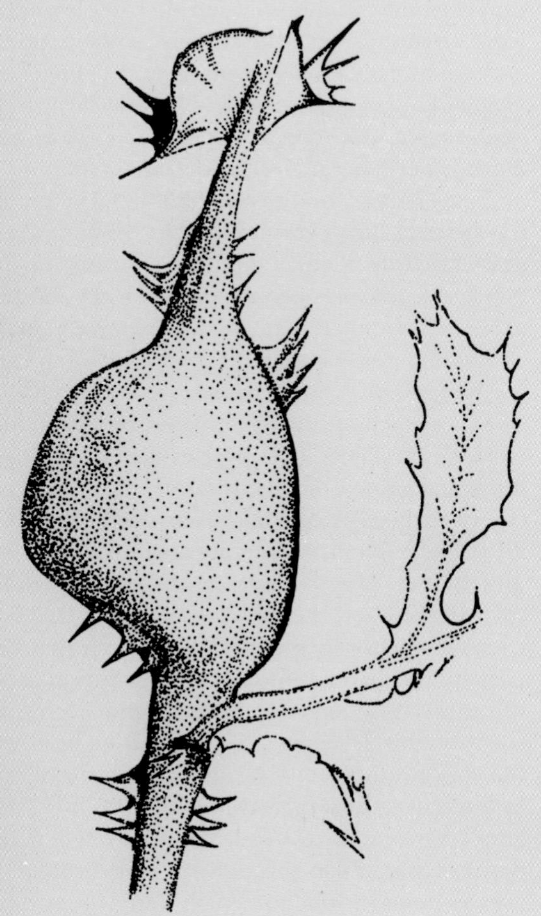

The creeping thistle fruitfly ***Euribia (Urophora) cardui*** *forms stem-galls on its foodplant.*

Many gall-wasps exhibit an alternation of generations. Each of these gives rise to both a different-looking adult insect and a very different-looking gall. The springtime generation of the very common European ***Neuroterus quercus-baccarum*** consists entirely of overwintered females. These lay unfertilized (parthenogenetic) eggs within the catkins of oaks, yielding red currant-like galls, producing wasps of both sexes by early summer. After mating in the normal way, the female now lays her eggs not in the catkins (which are long gone) but on the underside of the oak leaves. This results in flattened disc-like structures called spangle galls. These fall to the ground in autumn, complete with the wasp pupa, which emerges the following spring to begin the cycle again.

Not all galls are formed conspicuously out in the open. Many kinds occur on roots, where the causer can be a pest. The turnip gall-weevil **Ceutorhynchus pleurostigma lays** its eggs either on or close beside

The red bean-shaped gall of the sawfly ***Pontania proxima*** *is common on willows,* ***Salix*** *spp. The gall is often fully-formed before the sawfly's larva has even hatched, indicating that the chemical trigger for the gall's formation is introduced by the female at the time of oviposition* (left).

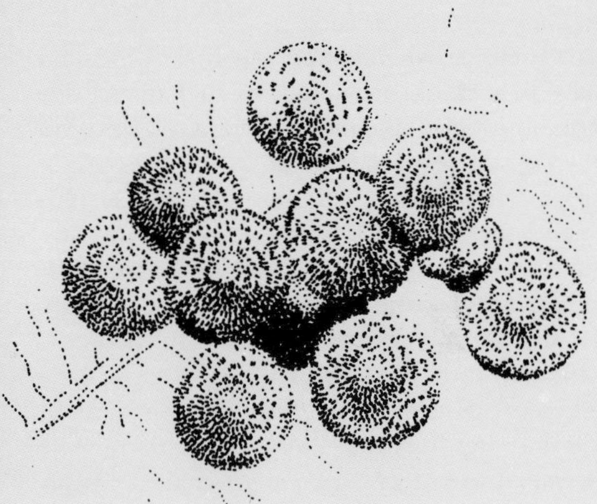

These spangle galls caused by the bisexual generations of the wasp ***Neuroterus quercus-baccarum*** *occur on the undersides of oak leaves. The unisexual generation gives rise to currant galls on the oak catkins.*

A knopper gall on acorns formed by the gall-wasp ***Andricus quercus-calicis.***

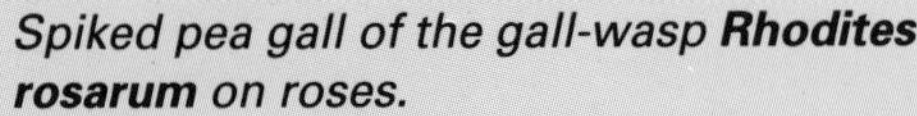

Spiked pea gall of the gall-wasp ***Rhodites rosarum*** *on roses.*

roots of brassicas such as turnips and cabbages. The gall produced can reach the size of a small marble, and when present in some numbers can seriously weaken the plant.

Many galls become a focus of attention for so-called inquilines, which occupy the galls alongside the rightful residents. These are also attacked by parasitic wasps.

people therefore realize that the aphids (*Aphis fabae*), which are such a pest on garden beans, have spent the winter as large black eggs on spindle trees (*Euonymus europaeus*).

Many small sap-sucking bugs, such as aphids, do not have to work to earn a meal by sucking actively from their host plant but merely sit and relax as food is forced up the feeding rostrum by the pressure of sap from active growth. The considerable surplus which thereby gains entry is passed out of the rear end as sugary honeydew. This in turn is eagerly lapped up by numerous ants which, jealous of a continued supply, defend the producers from predators. Indeed, some ants go so far as to farm their 'aphid cows', moving them up onto stems at night to feed and even building shelters over them. The yellow meadow ant (*Lasius flavus*) makes doubly sure of summertime honeydew deliveries by giving a home to aphid eggs during winter, taking them out onto the correct host plant when spring arrives.

INSECTS AND FUNGI

A number of insects make use of fungi as food and a large bracket fungus jutting forth from a tree trunk is likely to provide board and lodging for a menagerie of beetles and flies of several families. The often brightly coloured erotylid beetles are closely associated with fungi. *Pselaphacus giganteus* from the rainforests of South America actually herds her brood of wriggling babies from one developing fruit-body to another of *Polyporus tenuiculus*. This small glistening white bracket fungus is highly perishable, so faced with a staple having a short shelf-life, the larvae swarm over it in a non-stop day-and-night orgy of consumption. With a constant supply of food entering at one end, it is inevitable that an unending stream of slimy white liquid erupts from the other. This falls and contaminates any fungi occurring lower down but the mother is careful to shepherd her brood only to unsoiled specimens, dragging her rear end on the log to lay down a trail pheromone as guidance.

The larvae of many mycetophilid flies secrete silk from their mouthparts to lay as easily followed trails over their rather slimy fungal food. In the fungus-gnat (*Leptomorphus bifasciatus*), the larva fishes for its living by slinging a silken sheet beneath a bracket fungus. Showers of spores falling onto the net soon form a dense fungal

This throng of flies on a stinkhorn, **Phallus impudicus**, in an English forest has been attracted by the powerful stench of rotting flesh, emitted by the gluten which covers the head of the fungus. After eating the gluten, the flies broadcast the spores.

icing, which the larva regularly goes out and eats, net and all. The gap which results is soon made good using more silk from the mouthparts.

Many termites, along with the leaf-cutting ants, guarantee a more predictable food supply by cultivating their special domestic fungi in closely monitored gardens. *Atta* spp. ants cater for the needs of their fungus by raiding forest trees for supplies of leaf segments, snipped off with their mandibles. A steady flow of ants, each brandishing its fragment of greenery, trots down into the subterranean nests. On arrival, the leaves are chewed up, incorporated into compost in the 'garden' and quickly inoculated with a seed-corn of fresh fungus. In the warm humid confines of the nest, the fungus thrives, soon producing swellings which are the sole food of the ants. Interestingly enough, the particular species of fungus has never been found growing freely outside the nests of leaf-cutter ants. When a queen leaves the nest, she takes with her a pellet of the fungus in a special pouch beneath her mouthparts. Unlike in ambrosia beetles and others, which carry spores of their own particular fungus, the queen ant takes a speck of mycelium with which to found her new garden. Her first act upon ejecting the pellet is to provide it with manure, clear amber droplets from her anus, without which the fungus will not break into growth. In the absence of their custodians the fungus gardens soon become riddled with invasive fungi of various kinds, as well as bacteria and weeds. How the ants normally keep these so effectively at bay is still uncertain but is probably through a combination of chemical and physical means.

Flies such as bluebottles (*Calliphora* spp.) and greenbottles (*Lucilia* spp.), which are attracted to the stench of putrefaction, are easily lured to the stinking slimy black spore-containing gluten which covers the head of the stinkhorn fungus (*Phallus impudicus*). The flies avidly feed on the gluten, which soon instigates a bout of diarrhoea, thereby neatly ensuring that the spores pass rapidly through the flies' digestive system before incurring any damage. By the time this occurs, the fly will probably have flown some way from the original fungus, thus helping to disperse the spores, as intended.

INSECTS AND FLOWERS

The smell of rotting flesh is also mimicked by the flowers of a number of African plants in the milkweed genus *Stapelia* and its allies. Upon arrival, some flies are so completely fooled that they

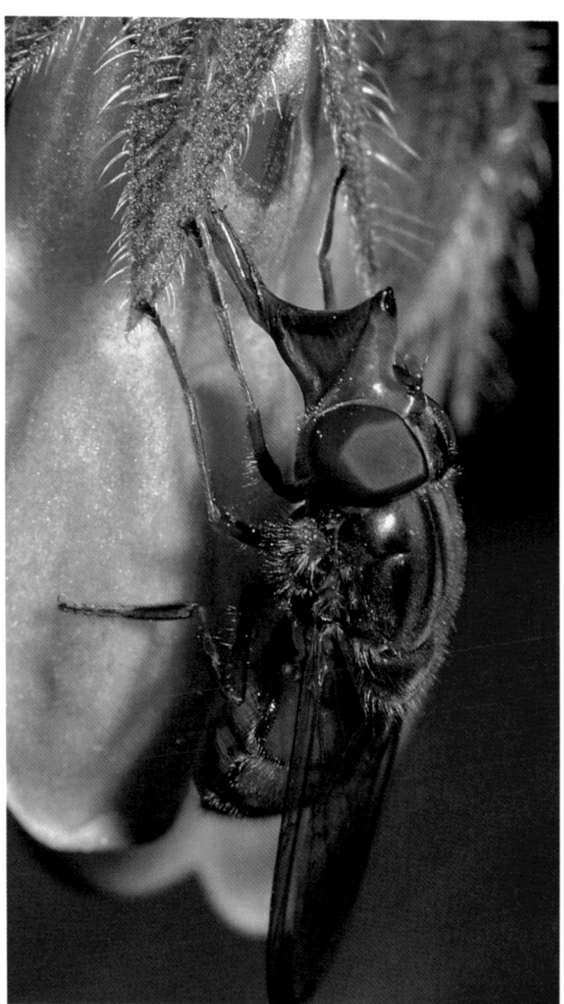

Larceny on an English roadside. A short-cut to the nectary of this comfrey flower, **Symphytum officinale**, has already been bitten out by a large bumble bee. All this **Rhingia campestris**, hoverfly (Syrphidae), has to do to rob the flower of its nectar is insert its proboscis through the ready-made hole.

While visiting flowers for food, beetles often become quite heavily contaminated with pollen. Unlike bees, they do not comb it off the body and into special baskets for carrying. This bee-wolf (**Trichodes ornatus bonnevillensis**) (Cleridae) in Utah will have spent its larval stages as a destructive parasite within the nest of a bee. The rather similar-looking European bee-wolf, **Trichodes apiarius,** can be a pest within hives of the honey bee, **Apis mellifera**.

squander a batch of eggs upon the barren surface of the flower, condemning the resulting larvae to wander around in a fruitless search for something edible, until they finally starve to death. Other flowers use different methods to dupe insects into making visits, which are unproductive for the insects but essential for the flowers. The South African orchid *Disa ferruginea,* which offers no reward, is nevertheless able to rely on just one butterfly, the nymphalid *Meneris tulbaghia,* for its pollination. The orchid manages this by mimicking flowers such as the red-hot poker (*Kniphofia uvaria,*) to which the butterfly does make profitable visits for nectar. Orchids of several genera, including *Ophrys* in Europe, dupe males of several species of bees and wasps into making profitless visits by mimicking the relevant females.

Most insect–flower relationships are, however, of mutual advantage, in a partnership which has evolved in tandem over millions of years. Different kinds of flowers offer varying rewards to different kinds of insects, depending on the suitability of their mouthparts for reaching the pollen or nectar. Flowers with very long thin tubes, such as the popular garden *Nicotiana,* tend to be specialized for visits by hawkmoths (or in diurnal flowers, by butterflies). The long thin lepidopteran proboscis is well able to plumb the depths of the tube, draining nectar which is out of reach for shorter-tongued bees or flies.

Moth-pollinated plants, which are typically white, tend to be nocturnal and produce a heavy far-reaching scent. Bee-pollinated plants often provide a miniature map on their floral faces to help the bee find the source of nectar as quickly and efficiently as possible. These guidemarks are in the best interests of both flower and bee. Both will wring maximum benefit from a rapid series of visits to as many flowers as possible. The handling time per flower is shorter in experienced bees than in callow ones and is shortest for simple flowers in which the nectar is exposed in an obvious place. More complex flowers, which only relinquish their riches to an insect capable of operating a special mechanism, take longer to learn. Garden snapdragons (*Antirrhinum* spp.) are a good example. Inexperienced bees have to make several ineffectual stabs at depressing the flower's rather stiff hinge, before they grasp the principle and poke their heads deep inside to reach the nectar. This mechanism ingeniously forestalls visits by small insects, which could steal the nectar without touching the flower's anthers or stigmas. Only insects which are big and powerful enough to operate the hinge can gain entry, effecting pollination as they do so. Flowers with similar difficult-to-open mechanisms, which are operated by large bumble bees, include monkshoods (*Aconitum* spp.) and columbines (*Aquilegia* spp.), both of which are common in gardens.

With well-filled pollen baskets on its hind legs, a bumble bee, **Bombus pennsylvanicus sonorus** visits a flower in the Arizona Desert, USA. Like most bees, bumble bees stop at regular intervals to comb pollen off the body and into the pollen-baskets.

In many drooping flowers, such as some species of *Solanum* and *Cassia*, it is pollen rather than nectar which is denied to the visitor unless it carries out the correct procedure. In this instance, the knack is to vibrate the flight muscles rapidly in a quick buzz, bringing a shower of pollen down onto the insect's underside. Here static electricity holds it in place until it can be combed off by the legs and stored on the pollen baskets. Once a bee has invested time in learning to operate a tricky mechanism, it will make sense to stick with that kind of flower, as long as it continues to offer sufficient pay-off in terms of nectar and pollen.

Unfortunately for the flowers, although the system benefits both partners, it incurs a disproportionate amount of work for just one – the insects. Not surprisingly, they have come up with numerous short-cuts to reduce their workload, none of which is in the interests of their static partners. In the case of common comfrey (*Symphytum officinale*) the doublecross is executed by short-tongued bumblebees such as *Bombus terrestris*, which bite a hole near the base of the flower-tube. This opens up a direct bypass to the nectar, dispensing with the laborious exercise of hanging upside-down to reach up inside the flower and probe for the nectary. The drawbacks for the flower are compound, as other bees and hoverflies quickly learn about the back-door shortcut, greatly reducing the number of legitimate front-door visits which result in pollination.

Most bees collect both nectar and pollen, temporarily storing the latter in pollen baskets on the hind legs (in most bees) or on a dense brush of hairs on the underside of the abdomen (for example, leafcutter bees). Many bees visit a very limited range of flowers. For example, *Tetralonia malvae* from the warmer parts of Europe, specializes on

Moths are mainly nocturnal feeders, using their long proboscis to obtain nectar from the flowers they visit. This is the North American subspecies of the striped or white-lined hawk moth **Hyles lineata** (Sphingidae) visiting an evening primrose flower at dusk in Utah.

mallows (*Malva* spp. and *Lavatera* spp.), attaching huge cumbersome-looking bundles of creamy pollen to the pollen baskets on its hind legs. *Perdita opuntiae* from the USA surfs her way across the broad swell of pollen on open offer in the prickly pear cactus flowers (*Opuntia* spp.), which are the only ones it ever visits. So bounteous is the supply, that a single flower will provide all the pollen she needs to stock a cell.

Butterflies generally visit flowers solely for the nectar reward, being unable either to collect or utilize pollen grains, which are too large to fit inside the slim proboscis. The exceptions are some South American heliconiines, which take advantage of the proteins available in pollen to extend their lifespans to an unusually protracted degree. The tip of the proboscis scrapes up pollen grains and then kneads them into a ball, along with nectar, into which some of the amino acids from the pollen eventually dissolve. The enriched nectar can then be imbibed in the normal way and the pollen husks discarded.

Pollen or nectar are not the invariable products marketed by flowers. Some are able to offer their services as oil-vendors. *Diascia longicornis* from South Africa promises ample stocks of nutritious oil but restricts the take-up by storing its supplies in a reservoir tucked away rather inaccessibly at the tip of a 25-mm long spur. The melittid bee, *Rediviva emdeorum*, is the only insect boasting the necessary hardware to liberate the supplies, having superlong front legs which are delved into the spurs and pick up the oil.

THE HUNTER-KILLERS

Insects have been making a living by killing other insects for a very long time. The dragonflies were the earliest practitioners of the art of aerial slaughter and they have been perfecting their slick technique for the last 350 million years. Dragonflies make the perfect aerial killing-machine. Their huge bulbous multifaceted eyes, set on highly mobile gun-turret heads, can acquire and follow fast-moving airborne targets at some considerable range. The bunch of six forward-pointing legs, armed with gin-trap spines, then comes into play to scoop up the prey. Just about anything which flies is likely to succumb, including *Vespula* spp. social wasps, which are apparently afforded scant chance to retaliate with their powerful stings. The meal is usually consumed on the wing, often accompanied by noisy crunching as indigestible items such as wings and legs are bitten away and rain earthwards. The most frequent items on the menu are flies and other small insects but occasionally a feast takes place when another dragonfly of only slightly inferior size falls victim. The giant damselfy, *Megaloprepus leucosticta*, from the tropical Americas, has evolved the surprising habit of plucking spiders from their webs.

Robberflies (Asilidae) also generally take prey on the wing, and may notch up kills of damselflies and even small dragonflies, as well as other robberflies, including members of their own species. They will also tackle bees and wasps, although the role of hunter and hunted may be reversed by hunting wasps such as *Bembix* and *Rubrica*, which specialize in taking flies, including large robberflies! Most robberflies are ground-to-air interceptors, perching on a stone or twig and arcing upwards to snatch their quarry out of the air. Death is almost instant, as the robberfly's sharp proboscis punctures the delicate membrane between the victim's head and thorax. If the prey does manage to put up a brief struggle before it succumbs, the robberfly's eyes are protected from desperately kicking legs by a prominent 'moustache' of downward-pointing hairs.

Hangingflies are not related to robberflies, being members of the order Mecoptera. They employ a very different hunting technique, effectively lassoing their prey with their long back legs. These hang down in flight and are stroked across vegetation, coiling around and hauling up any suitable target. The hunter probably never makes visual contact with its prey but detects it solely by tactile feedback from the sensitive back legs as they trawl along.

Wasps, both solitary and social, are among the most businesslike of killers. The most impressive exponents of the art of clinically efficient execution are the solitary wasps which specialize on dangerous prey, such as spiders, bees, wasps and mantids. The metallic blue-grey *Pepsis* wasps are obliged to come to grips with tarantulas larger than themselves. At first sight this seems a suicidal prospect, but in reality, the spiders seldom put up much of a struggle, almost passively awaiting their fate and allowing the wasp to creep in and sting them. In fact, blind panic, rather than outright life-saving aggression, seems to be the reaction in most spiders confronted by the vibrating antennae of a hunting wasp closing in for the kill. The wasp does not hesitate to meet the spider on its home ground by entering its burrow. Yet, instead of encountering the hostile reception reserved for most intruders, the wasp is likely to find the lair empty, its occupant having made a bolt for it through an emergency exit. Once outside, the spider may yet

Hunting-wasps which specialize in taking grasshoppers have to beware of the prey's defences. These include biting jaws able to inflict mortal wounds, and the ejection of a sticky and offensive liquid from the mouth. This **Sphex** sp. wasp is dragging a large grasshopper to her burrow in a Kenyan forest. The provision of such a large single meal will suffice for the complete development of the wasp's larva.

Efferia benedicti is one of the commonest robberflies (Asilidae) in south-western USA. This one made a lightning dart upwards to capture a sulphur butterfly, **Colias eurytheme** in Utah. **Efferia** is one of the largest genera of robberflies in the USA, with over 105 described species.

Most pompilid wasps dig a shaft in which to entomb the spider before laying an egg on it. The North American *Anoplius cylindricus* is one of a number which saves time by merely digging a side extension to the spider's own shaft and putting it in there. Another North American species, *Anoplius depressipes*, exhibits remarkable tenacity in prosecuting its chase of the swamp spider, *Dolomedes triton*. When the thoroughly intimidated spider plunges below the water in a bid to escape, the wasp dives in after it and is capable of stinging it while submerged and dragging it back to the surface.

Wasps which hunt crickets run the gauntlet of taking a hefty kick in the face from the powerful back legs. This can send the attacker spinning off for quite a distance, while the spines on the cricket's legs are capable of causing considerable damage to its opponent's wings or antennae. The wasp therefore tries to minimize the hazards by plunging home its first sting in the large nerve ganglion which serves the powerful leg muscles. For wasps such as *Tachysphex* and *Prionyx* which take grasshoppers, it is the prey's mouthparts which pose the greatest threat. Not only are these capable of administering a fatal bite if the wasp positions herself wrongly during the struggle; they can also spew forth a deluge of odious gut contents. This tends to flow down across the grasshopper's underside, inhibiting the wasp in placing an egg because of the need to avoid contaminating the body surface with such a sticky and repellent fluid. The best tactic therefore is to inca-

escape with its life if it lies doggo, while its tormentor flies around in a frantic search. It seems that to many wasps the burrow, betrayed by its pheromone-impregnated silk, is easier to find than a lone spider cowering motionless in the open.

The **Chlorion lobatum** solitary wasp, stinging a cricket in Nepal, has to be careful of its prey's powerful jaws and heavily spined back legs. In fact this cricket escaped after letting fly with a massive kick.

THE UNDERWATER KILLER

Wasps and other insects which land on the surface of a pond to slake their thirst run the risk of being speared on the beak-like rostrum of a backswimmer bug (*Notonecta* spp.). Death follows swiftly, for the beak injects a powerful paralysing poison, which is extremely painful when injected into a human finger. Falling victim to a backswimmer is more typically the fate of an egg-laying damselfly, despite the best efforts of her mate to save her.

pacitate the troublesome jaws right at the start by trying to pinpoint the first sting in the neck region.

Wasps such as *Palarus* and *Philanthus* which specialize in taking bees may also have a battle on their hands before the victim is subdued. However, once this is secured, the wasp can refresh itself after its exertions by turning its victim's head around and squeezing a snack of nectar from its crop.

Social wasps such as *Vespula* spp., the troublesome yellowjackets of warm summer days, need to locate a constant supply of fresh meat to satisfy the multiple appetites of the larvae back in the nest. The time invested in searching is enormous, and after a period of lean pickings a bootyless wasp will even try its luck in plucking a plump spider from the centre of its orb-web. A tactic which tends to be more auspicious is to loiter near a group of flowers. While the attendant bees and hoverflies are fully occupied in the pursuit of nectar, they make tempting targets for a quick pounce by the waiting wasp. Although there are more misses than hits, there are still plenty of occasions when the wasp manages to establish a firm grip with its legs and jaws. The ensuing struggle usually precipitates both contestants onto the ground, where the wasp is at an advantage in bringing its meat-cleaving jaws to bear, biting off the victim's head. The ensuing job of butchering is performed quickly in a series of deft-practised actions. Waste items such as legs and wings are snipped off, before the meaty abdomen and thorax are folded up into a compact package suitable for ferrying back to the nest. Caterpillars are also frequent targets, even quite hairy ones, which are often shunned by solitary hunting wasps. Social wasps are valiant antagonists, unafraid to mount a frontal assault against an assassin bug poised ready in attack mode, with legs raised. The intensity of their labours leads to frequent need for refreshment, usually from flowers or by landing on the surface of a pond or stream. In times of drought, when free water is rare, the wasps demonstrate their amazing adaptability by creating their own drinking-fountains. By chewing off flowerheads such as creeping thistle (*Cirsium dissectum*), they create a sap flow which satisfies not only their own thirst but also that of other insects such as ladybirds, which follow on behind.

All the examples quoted so far involve adaptations for the predatory way of life. These include sharp eyesight, a high degree of aerial combat maneouvrability, large jaws or a powerful sting. The staphylinid beetle, *Stenus bipunctatus*, even has a chameleon-style extensible labium which snaps up its prey from long range and hauls it back to the mouth. Another beetle, the European carabid, *Loricera pilicornis*, pre-empts a quick escape-jump from its springtail prey by trapping it in a cage formed by hairs at the base of the beetle's antennae and head.

Some insects conspicuously lack any modifications for their predatory lifestyle, simply because they feast on prey which is so abundant and passive that it merely sits and waits to become the next meal. Ladybirds, for example, can get away with being rather cumbersome without sacrificing their ability to secure a steady supply of their aphid food. Slug-like hoverfly larvae cannot even see the aphids which surround them but all they have to do to select the next morsel is wave their head around and they are bound to make contact sooner or later. They are assisted by the fact that the aphids do little to avoid capture. Those lacewing larvae which feed on aphids do at least have proper legs and respectably large sickle-shaped jaws, put to good use in spearing a soft-bodied victim and holding it aloft while, kicking helplessly, it is drained dry. In a departure from the normal routine of taking easy pickings from the midst of an aphid colony, the North American mirid bug, *Ranzovius contubernalis*, obtains its aphids second-hand in a most remarkable place, the webs of two species of spiders. The bug moves unhindered around the webs, feeding on trapped small insects, mostly aphids, which fall below the size threshold of interest to the spider.

SIT-AND-WAIT KILLERS

Rather than expending energy in seeking out prey, many predatory insects take a more relaxed approach and wait for a meal to arrive under its own power. Praying mantids are the perfect exponents of this, sitting quietly until the vice-like front legs can safely be raked forwards, taking an incautious intruder by surprise in a fatal embrace. Many mantids are simply green or brown, making them inconspicuous on leaves or bark. Some species, such as the nymphs of *Pseudocreobotra ocellata* from Africa, blend perfectly into pink or white flowers, or even mimic a flower in its own right, as does the African Devil mantis, *Idolium diabolicum*. Mantids do not always dispatch their captives quickly and relatively painlessly, as do most wasps and robberflies. Instead, they tend to set about satisfying their hunger the instant their victim is securely clamped in the front legs. With little chance of escape, the prey must often suffer being eaten alive as the mantis begins to browse off strips of flesh with gently nibbling mouthparts. When the sufferer is a vertebrate animal such as a frog, its shrill cries are a piteous reminder that insects have no feelings of compassion for their victims.

Adult mantispids have similarly raptorial front legs but whereas praying mantis nymphs catch prey much in the manner of the adults, mantispid larvae lead very different lives. The tiny larva of *Mantispa uhleri* waits for a female of the wolf spider (*Lycosa rabida*) to pass by and then jumps on board. It then sups on the blood of its unwitting host until the spider lays its eggs, whereupon the mantispid larva bores into the egg-sac and consumes its contents.

Many assassin-bugs (Reduviidae) are ambush specialists, lying in wait on leaves, bark or flowers until something turns up. However, a few kinds have developed some remarkably exotic methods of capturing prey. The nymph of *Salyavata variegata* spends its life pacing slowly around on the arboreal carton nests of *Nasutitermes* spp. termites in Costa Rica. When the worker termites make a small opening in the nest's protective outer coating, in order to extend it, they unwittingly give the waiting bug the opportunity it has been waiting for. It dips its long stiletto-like rostrum into the opening and spears a worker. Within minutes, this has been sucked dry, whereupon the bug uses its deflated corpse to 'fish' for more termites, proffering it as bait on the tip of its rostrum. This system works because the termites are strongly attracted to their own dead, which are recycled as food.

Pondskaters or water-striders are quick to detect the ripples generated by an insect struggling on the surface film. These toothed pondskaters, **Gerris odontogaster**, in England are feeding on a damselfly, using their piercing mouthparts to inject enzymes which liquify the prey's tissues. Although the legs of a pondskater are very thin, they do not penetrate the meniscus because they are covered with unwettable hairs.

Bait of a very different kind figure in the strategy employed by another assassin-bug, *Amulius malayus*, from South-east Asia. It swabs resin off a tree with its hairy front legs, and then uses it to coax *Trigona* spp. stingless bees to fly into its arms,

INSECTS WITH FISHING LINES

A number of mycetophilid fungus-gnats have, in their larval stages, borrowed the habits of spiders in building a kind of web to ensnare flying insects. In several cases, such as the Australian *Arachnocampa flava*, the waiting gnat larva encourages passers-by to head towards the sticky droplets on its 'fishing lines' by shining a light, both larva and web being highly luminous. After nightfall, certain damp cliff faces in the Queensland rainforest are brightly spangled with hundreds of these webs, glowing like miniature fairylights.

lured by the resin, which they use as a nesting material. Females in the New World firefly genus *Photuris* use light as a bait. From a leafy perch, they send out a coded sequence of flashes which copies those emitted by the females of several other species of fireflies, mainly in the genus *Photinus.* Duped by this counterfeit invitation to a sexual rendezvous, passing *Photinus* males drop in, only to lose their heads rather than their virginity.

Using bait is not very far removed technically from laying a trap, of which several kinds are used by different insects. The pitfall traps laboriously prepared by ant-lion larvae in soft earth or sand resemble miniature craters. The owner lurks just beneath the sand at the bottom of its pit, directing a bombardment of grains straight at any insect which wanders over the edge. With luck, this will send the target tumbling down towards the larva's capacious jaws. Similar pitfall traps are constructed by the so-called worm-lions, fly larvae belonging to the subfamily Vermileoninae within the family Rhagionidae.

SCAVENGERS AND KLEPTOPARASITES

There is no shortage of insects to take on the necessary job of recycling waste material, such as dung or corpses. The useful task performed by dung beetles has already been mentioned in Chapter 5. If they fail to arrive quickly enough, especially in Africa, then they could find access to their supplies blocked by a dense struggling mass of butterflies, especially if the dung has come from a predator, such as a cat. Dung, urine, suppurating corpses, rotting fruit, fermenting tree sap or salty ground are all powerful attractants for male butterflies. Regular visits to such products are thought to be necessary for replacing nitrogenous salts lost during mating with numerous females. 'Puddling' behaviour on damp salty ground often gives rise to huge aggregations of butterflies on tropical, and occasionally temperate, riverbanks. A number of

Salt-seeking 'puddling' behaviour by adult butterflies and moths on damp paths and riverbanks is commonly seen throughout the tropics, and to a lesser extent in temperate areas. With its conspicuously 'tailed' hindwings, this day-flying moth, **Urania leilus** (Uraniidae) from Peru is deceptively similar to a swallowtail butterfly (Papilionidae).

butterflies and moths have switched to procuring their salt from the tear-ducts of various animals, including caymans, cattle and even man. Some ithomiine butterflies from South America, such as *Mechanitis polymnia*, are known as ant-butterflies from their habit of following raiding columns of *Eciton* spp. army ants. It is not the ants themselves which are the magnet, but the droppings left behind by ant-birds, which make a living by snapping up insects fleeing from the raiding swarms.

Panorpa spp. scorpionflies are adept at trespassing in spiders' webs in order to steal the dead insects trapped therein. Although obliged to move around the web with extreme caution, taking care not to alert its owner, the scorpionflies do not seem to have any difficulty in negotiating the actual silken lines. If something goes wrong and they do get stuck, they can deploy a quick-release mechanism in the form of a brown liquid, which they squirt from the mouth, dissolving the entrapping silk. Scorpionflies will feed upon just about anything which is dead, including mammal corpses heavily stained with blood and partly sucked insect prey dropped by robberflies. A favourite dish is the plump corpse of a fleshfly (*Sarcophaga* sp.) or other fly stuffed to bursting with the spores of the entomophagous fungus *Entomophthora musci*. These are common in damp summers, when the fungus is particularly virulent. The scorpionfly can eat its fill with impunity, because the fungal spores will only germinate in the body of true flies of certain genera.

In the tropics, male butterflies can be attracted in some numbers to fresh dung, especially when it originated from a predator, with leopard being reckoned as the best. This is **Charaxes brutus** in a Kenyan rainforest.

The strategy of habitually showing up as an uninvited guest at a superior's table might seem risky but is practised by a number of insects. The safe-conduct pass in such a situation is to be so small that the predator does not take any notice. Thus tiny milichiid flies often string along with spiders, assassin bugs or powerfully built robberflies in order to sup up the liquids oozing from the lacerated bodies of their prey. The diminutive pilferers walk with impunity across the surface of their benefactors, being an irritation at most. The carabid beetles, *Helluomorphoides latitarsis* and *H. ferrugineus*, set themselves up as highwaymen on the raiding trails of the ant *Neivamyrmex nigrescens* in Arizona. This is not as hazardous as it seems, because the ants just seem to drop their prey as soon as they encounter the beetle, without so much as a perfunctory show of resistance. They even allow the beetles to mug them by tugging the prey from their very jaws. If they chance upon a nest, the beetles will turn on the ant larvae, gorging themselves to the point of bursting their overstretched abdomens. Nests of both ants and termites are popular targets for resident scroungers, especially beetles, which manage to obtain regular meals by correctly mimicking their host's food-begging behaviour.

THE BLOOD-SUCKERS

Insects which take blood from a living mammal are of particular interest to humans. Not only are the bites of mosquitoes, tsetse flies and midges an irritation, they may also pass on potentially fatal diseases such as malaria and sleeping sickness. The proboscis functions not only as an inlet tube but also as a delivery pipe for a dose of anticoagulant, which prevents rapid blockage of the narrow feeding-duct. In most cases blood-sucking is restricted to the females; the males visit flowers for nectar. The biting midges (Ceratopogonidae) do not confine their attentions to warm-blooded animals such as man but also attack frogs and, to a larger extent, other insects. These tiny midges often attach themselves to the wings of bigger insects such as lacewings or dragonflies, and also suck the blood of large caterpillars. One species *Culicoides anophelis* even pursues mosquitoes distended with mammalian blood and drains them of a little for its own use by puncturing the abdomen. Perhaps the most unusual exponent of the vampire's art is the Asian noctuid moth *Calpe eustrigata*, although several of its close relatives pierce fruits rather than the skin of man or cattle.

Snipe flies of the family Rhagionidae are usually thought of as harmless creatures who, in the British Isles at least, just sit around on tree-trunks doing nothing in particular. In the mountains of the American southwest, however, a number of more sinister members of the family in the genus **Symphoromyia** are blood-suckers. Shown here taking a meal from the brave wife of one of the authors.

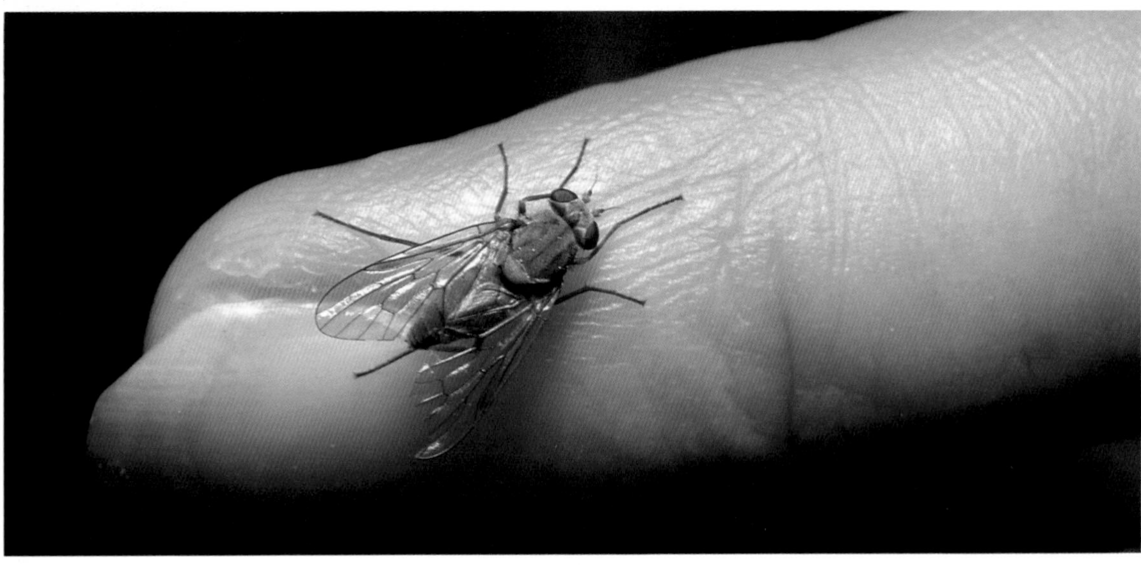

07 Defence

Most insects are small enough to make it highly likely that they are going to end up as food for some other animal. Spiders probably make the greatest inroads into insect numbers, while other insects probably come a close second, with frogs, toads, lizards, birds and mammals also taking their toll. Even the largest of insects, such as bulky armoured goliath beetles or 20cm long spiny-legged praying mantids may meet their match when confronted by powerful suitably equipped predators such as hornbills or monkeys. Fortunately, insects are able to fight back in a myriad different ways, not least of which is their frequent ability to overwhelm their enemies by breeding rapidly in large numbers.

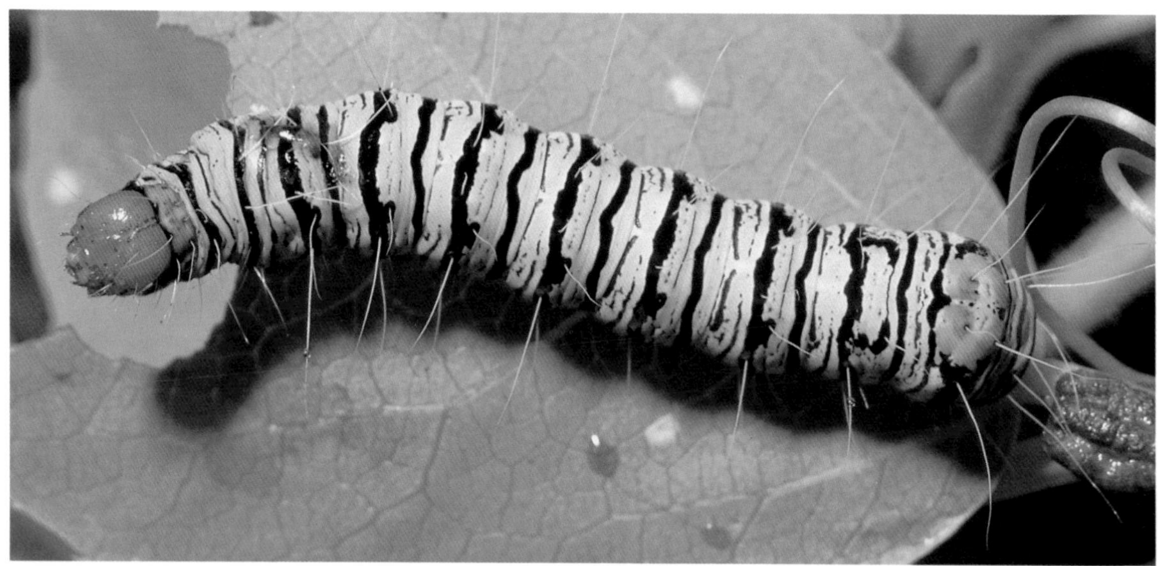

This moth caterpillar in South Africa has reacted to disturbance by vomiting up a gob of disagreeable liquid, visible as yellowish smears behind the head and on the leaf. As a back-up defence, the caterpillar also sports a bright orange 'false head' at its rear end, designed to draw a predator's attack away from the real head. Such 'false-head mimicry' is common in both larval and adult Lepidopterans. Among adults, it is particularly evident in many 'blues' (Lycaenidae) in which eye-like markings near the tips of the hindwings are combined with slender appendages which mimic antennae. Up-and-down shuffling movements of the wings lend a life-like movement to the 'antennae', drawing the maximum amount of attention to the 'false head'.

PHYSICAL DEFENCES AND ESCAPE TACTICS

DEFENSIVE WEAPONS

Many large katydids (Tettigoniidae) and longhorn beetles (Cerambycidae) can easily draw blood from a human finger with their powerful jaws, making it essential to handle them with extreme care. The same caution applies to assassin-bugs (Reduviidae) and backswimmers (Notonectidae), both of which can stab vigorously with a rigid rostrum and inject a

AGGRESSIVE CATERPILLARS

Many caterpillars thrash wildly around when touched, often combining this with vomiting up a gob of liquid derived from their food. This is held in readiness on the mouthparts and then wiped across the attacker's skin or mouth. This liquid can be very objectionable, containing in concentrated form whichever unpleasant chemicals the caterpillar has taken in from its foodplant. Many grasshoppers adopt the same tactic.

Hairy caterpillars often form dense defensive aggregations. These **Malacosoma californicum**, California tent-caterpillars (Lasiocampidae), are grouped on a silken web (the tent) which they have built on a cottonwood tree in a desert canyon in the state of Utah, USA. The related American tent-caterpillar, **M. americanum**, is known to lay down pheromonal recruitment trails between the tent and the best feeding areas.

The underwater environment is rather an unusual one for the caterpillar of a moth, but this is just where the larva of the chinamark moth, **Nymphula stratiotata**, is at home. It lives inside a portable case built from semi-circular sections bitten from pondweed leaves, and can withdraw inside if danger threatens. The case is superficially similar to some made by unrelated caddis-fly larvae (Trichoptera). The caterpillar's body is fully exposed to the water, with no bubble of air trapped inside the case, and it appears that breathing is performed by direct absorption of oxygen from the water itself.

very painful poison. The exposed face of an opponent, especially the delicate tongue, palate and eyes, or the vulnerable nose and lips of a monkey, are especially sensitive to the prodding defences employed by many large katydids and grasshoppers, which have powerful stabbing back legs armed with rows of backward-pointing spines. These are held up in readiness in a 'cocked' position, giving advance-notice of what is to follow if an enemy presses home its attack. If the insect is eaten whole, the spines may make the predator vomit up its meal within a few minutes, making it think twice next time. With mantids it is the heavily spined front legs which are splayed out wide, poised ready to come raking painfully down across exposed skin or nasal membranes. In the longhorn beetles of the genus *Hammaticherus*, the long antennae bear projecting spines. If a hand (or beak, with exposed eyes close-by) descends onto the beetle's back, it instantly lashes the antennae rearwards and then waggles its head, pulling the antennal hooks deeply into the assailant's skin. The pupae of many beetles, especially ladybirds (Coccinellidae), are provided with a gin-trap mechanism. If disturbed, most likely by an ant, the pupa snaps downwards, possibly nipping an antenna or leg of the assailant, although the action itself also seems to be a powerful deterrent.

ARMOUR-PLATING, SPINES, THORNS AND SHIELDS

Armour-plate protecting some or all of the body is found in many katydids, beetles, weevils and some bugs. It may be reinforced with ramparts of rigid spines projecting from the head, thorax and legs.

This makes the insect difficult to deal with except for predators with powerful crunching jaws, like a mongoose or civet. Bats, with their rather flimsy easily-punctured wing-membranes, will not even attempt to touch a really ferociously spined katydid, detecting its armament by echo-location.

The thorn-like outgrowths on membracid bugs such as *Umbonia spinosa* are also more a means of making them difficult to swallow than actually mimicry of thorns. At close range the 'thorn bugs' are in fact most unlike the genuine article and vastly larger than any real thorns, which seldom occur on the host trees. Some tortoise-beetles present a smooth shiny shield-like exterior to the outside world, making them immune to attack by smaller predators such as ants, as long as the vulnerable underside can be protected. To prevent itself being capsized, the beetle clamps itself tightly against the leaf on sucker-like feet.

HAIRS

The dense mass of hairs which clothes many moth and butterfly caterpillars is also effective in deterring most enemies from taking a bite. As always, there are exceptions, such as the European cuckoo which is adept at stripping the offensive skin from the edible interior. The hairs are often brittle and designed to break off in the attacker's skin. Once in place, they continue to deliver an irritant poison, which can cause painful urticating symptoms over a very long period. As it spins its cocoon, the caterpillar often incorporates some of its urticating hairs into the silk, providing extra protection when it becomes a pupa. Some female moths also screen their eggs behind a mass of hairs scraped off an anal tuft, which contains glands secreting irritant material.

HOUSES AND SECRETIONS

A portable blockhouse into which its owner can retreat at times of threat is the principle behind the cumbersome structures carried by the larvae of bagworm moths (Psychidae). The basic framework consists of silk, into which bits of twig and leaf are woven. The larva needs only to pop its head back inside and pull the mouth shut behind it. Similar portable redoubts, incorporating such natural construction materials as snail shells (sometimes with the living owner still inside), pebbles, bits of pondweed and fallen pine-needles are also typical of most aquatic caddis-fly larvae.

Tortoise-beetle larvae generally employ a shield made of their own droppings, often combined with their shed skins and hairs grazed off their food plant. These are held on a forked hinged tail-process which can be manipulated at various angles to parry an enemy's attack. Ants, in particular, are repelled by the droppings. Many insects, especially true bugs, but also some beetles and caterpillars, secrete large amounts of disagreeable glistening white wax. This usually projects in stringy filaments from glands at the rear end. In large adult bugs such as *Lystra lanata* (Fulgoridae), it trails out behind as a long waxy slipstream during flight. Second-hand supplies of wax are acquired by the larva of the green lacewing, *Ceraeochrysa cincta.* It tugs the waxy secretions off its whitefly prey with its mouth and then sticks them on its own back. This provides both a visual and chemical screen against recognition by ants and other enemies.

ESCAPE TACTICS

The simplest reaction to an imminent threat is to jump or fly out of danger. Weevils generally do neither of these – instead they simply tuck their legs in and dive head-first off a leaf or stem. They then play dead wherever they chance to land, being extremely difficult to find if this happens to be the ground, which it usually is. Stick grasshoppers (Proscopiidae) tend to do likewise, there generally being an abundance of sticks littering the ground among which to disappear quietly. The exceptions are the few species which are outstandingly good mimics of a specific food plant, in which case it is a safer bet to stay put. Many caterpillars respond to danger by launching themselves into space on a thread spun from the mouthparts. They eventually haul themselves back up, eating the silken line as they go. Unfortunately for the caterpillar of the green cloverworm moth, the parasitic wasp, *Diolcogaster facetosa* has the measure of this ploy, diligently following the thread down so that there is no escape for the caterpillar.

CHEMICAL DEFENCES

Insects are the prime exponents of chemical warfare. The most famous is the bombardier beetle, which can blast an enemy full in the face with a spray of quinones emitted from an internal discharge chamber which reaches a temperature of 100°C. To achieve this the beetle's own body acts as a kind of retort in which various chemicals are mixed, producing the explosive discharge through a directable nozzle at the tip of the abdomen. Ants which receive a direct hit are disabled and reduced to drunkenly staggering wrecks for several minutes. Any fellow ants which were not in the direct line of fire are dissuaded from further approaches by the quinones, which remain plastered over the beetle's exterior. The beetle can produce quite a cannonade, being able to maintain a rapid-fire barrage of up to twenty rounds in succession, each one accompanied by an audible pop and a puff of smoke.

Most examples of the application of chemistry in the insect world are less dramatic, although they can be no less traumatic to the animal on the receiving end. In their development of an extensive chemical armoury, the insects have been greatly (and unwillingly) aided by the plants upon which they prey. In their ongoing struggle to protect themselves effectively against the insect siege, plants have come up with a veritable pharmacy of different drugs, designed to repel or poison their enemies. Unfortunately, the sheer adaptability so typical of insects has always kept them one step ahead. In no single component is this more emphatic than the way in which the insects have turned to their own defensive account the very substances originally intended to destroy or deter them. Either in their pure form, or modified within their own bodies and combined with self-generated compounds, the insects now use these often complex substances as highly effective weapons.

The waxy filaments trailing from the rear end of this Peruvian **Lystra lanata** bug (Fulgoridae) are found in a number of different genera and are thought to be defensive in function.

Despite containing relatively massive amounts of such lethal chemicals as hydrogen cyanide, this burnet moth caterpillar, **Zygaena lonicerae**, in an English meadow has fallen victim to a nymph of the shieldbug, **Picromerus bidens**. This particular bug will feed on burnet moths in their larval, pupal and adult stages, apparently being immune to any ill effects. Stinkbugs also feed with equanimity on other chemically 'protected' insects, such as ladybirds.

These are mainly deployed against vertebrate animals such as birds or mammals, which display greater sensitivity to disagreeable tastes, or succumb easily to the effects of poisons such as cardiac glycosides. These have a dramatic effect on the vertebrate heart but are of little worth against many arthropod predators, such as spiders or stinkbugs. However, even these vary in their resistance to defensive compounds. The stinkbug, *Picromerus bidens*, will feed on burnet moths in their larval, pupal and adult stages, despite the high amounts of poisons present, including hydrogen cyanide sequestered from the cyanogenic larval food plants, vetches and clovers. By contrast, dragonflies are but one example of a predatory arthropod which will drop a burnet moth as soon as contact is made.

The most harrowing and painful example of chemical warfare is in its injectable form, as evinced by the powerful stings of bees, wasps and some ants. Many of the latter augment the pain already inflicted by launching a follow-up attack from the jaws. The combined assault from both ends mounted by the pugnacious Australian *Myrmecia* ants has earned them the name of bulldog ants along with a fearsome reputation. Even relatively harmless-looking insects such as brightly coloured impassive slug-moth caterpillars (Limacodidae) can mount a powerful reprisal without even stirring a muscle. The merest brush with a finger across the tufts of spines which beset these squat creatures sends a searing pain lancing through the hand. In some south American

Members of the large worldwide moth family Arctiidae tend to be day-flying and brightly coloured, reflecting their possession of some extremely unpleasant defensive chemicals. This is **Gymnelia ethodaea**, from Mexico, one of many examples in the subfamily Ctenuchiinae in which transparent sections in the wings give them some resemblance to wasps.

Chemically protected insects typically flaunt a uniform of bright easily memorized colours. This is a mating pair of **Dactylotum bicolor** (painted grasshoppers) in the Arizona desert (*below*).

DANGEROUS STICK INSECTS

Irritant chemical sprays can also have a dramatically distressing effect upon the recipient, to which the author can amply testify. A pair of plump stockily built *Agathemera* stick insects (Phasmidae) resting quietly on a plant in the Patagonian desert seemed a harmless enough subject for a close-up picture. Yet seconds after moving in for a closer look, the author was reeling backwards, hands over eyes in an agony of surprise and discomfort. With eyes seemingly on fire, it felt as though someone had sprayed them with acid, and indeed, they had. The stick insects, still resting imperturbably in the same spot, had responded to a pair of mammalian eyes coming in much too close by squirting them neatly and accurately with an irritant chemical, released from nozzles in the thorax. The spray is invisible but its release is accompanied by a distinct hissing-snapping sound. With storage space being limited, the phasmids do not squander precious reserves by spraying an intruding (and insensitive) hand. Nor do they bother with a face which is out of range but only let fly, with perfect aim, at a pair of eyes within safe reach of their weaponry. Unable to leave the scene of its encounter with much haste, the wingless phasmid must rely on the ten minutes of blindness and distress induced by its spray to deter further aggression. Indeed, it is unlikely that any enemy, even the hungry-looking grey foxes, which haunt the Patagonian deserts, would be desperate enough to come back for a second go.

This pair of plump **Agathemera** sp., stick insects, in Argentina temporarily blinded the photographer with an irritant spray directed accurately into his eyes.

species, the caterpillar resembles a perfectly innocuous crinkled dead leaf, upon which the neat rosettes of short stinging spines are scarcely noticed. Not, that is, until the merest brush with bare skin sends an excruciating message that a second encounter would not be wise.

Chemical spray-gun defences are employed by stick insects (see box), and by the North African pyrgomorph grasshopper, *Poecilocerus hierogylphicus*. This can squirt a torrent of whitish liquid which tastes and smells revolting, and is disastrous when received in the eyes. Many darkling beetles (Tenebrionidae) can also emit a jet of unpleasant quinones from their rear ends. In the deserts of the southwest USA it is common to see a hurrying *Eleodes* sp. stop in its tracks and perform a headstand. Pointing the tip of its abdomen upwards towards its enemy, it gives an unmistakable visual signal that it is primed and ready to fire. A large number of insects can only expel their defensive chemicals as far as the surface of their own body, where it evaporates to form an odorous protective halo. Effective deterrence is sometimes earned only after a close investigation by an enemy's beak or nose. However, in some instances the stench permeates a broad zone, especially when released by an assembly of insects. When the author encountered large numbers of the cryptic stinkbug, *Coriplatus depressus*, on a tree-trunk in a Peruvian rainforest, it was the stomach-turning chemical miasma, strongly discernible from more than a metre away, that first betrayed their presence. Adult bugs liberate their defensive chemicals from pores on the thorax, whereas nymphs effect release from several pores, often clearly visible, situated on the top of the abdomen. In the lesser backswimmer, *Plea minutissima*, an aquatic bug, the output from these glands has been directed to a new use. These bugs rely on their own air supply carried in a bubble attached to special hydrofuge hairs on their undersides. If these hairs become contaminated with a growth of bacteria, the bubble collapses and the backswimmer drowns. Regular trips out of water are therefore made in order to smear the products of the thoracic glands (mainly hydrogen peroxide) across the underside, where its powerfully antiseptic action prevents fouling.

The South African milkweed grasshopper, *Phymateus morbillosus*, is one of many insects which feed upon poisonous milkweeds (Asclepiadaceae). These contain a cornucopia of unpleasant chemicals, the most important being cardiac glycoside heart poisons. If sufficiently tormented, the milkweed grasshopper expels a disgusting frothy liquid from vents near the base of the hindlegs. Another African species, *Dictyophorus spumans*, has even earned itself the name of foaming grasshopper on account of its copious output. The American lubber grasshopper, *Romalea microptera*, backs up a similar performance by emitting a threatening hiss. In the unlikely event that the noisome reek does not procure sufficient discouragement, the deterrent effect is increased with the appearance of an especially nauseating gob of liquid from the mouth.

If they are roughly handled – an exploratory peck would suffice – African acraeine butterflies can produce large amounts of protective foam from their thoracic glands. In *Acraea encedon*, one component of this foam is known to be hydrogen cyanide, which is also an element in the defences of other insects, including burnet moths (Zygaenidae). Many rove beetles (Staphylinidae) can discharge an evil-smelling liquid from the anal glands. The 'tail' is directable like that of a scorpion, so the fluid can be deposited accurately on an attacker approaching from virtually any angle, an attribute which is a special bonus against highly mobile ants. Lacewing larvae exercise similar tactics, taking advantage of the flexibility of the abdomen to plop a blob of anal liquid onto an assailant.

In a number of insects, it is the blood itself which is caustic or distasteful and takes over the defensive role. Unfortunately, the only practical means of getting the 'reject me' message across is to leak

The European bloody-nosed beetle, **Timarcha tenebricosa**, has to suffer considerable provocation before it will invoke its reflex-bleeding defensive performance.

some blood onto the body surface, where it can be sniffed or tasted. Any blood lost in this way has to be replaced, at some cost, so exponents of reflex bleeding, such as the bloody-nosed beetles (*Timarcha* spp.), oil-beetles (for example *Meloe* spp.) and ladybirds tend to be fairly well armoured. Only after a considerable amount of heavy-handed treatment will they finally part with some of their vital liquid reserves. Bloody-nosed beetles make the discharge of their bright red evil-tasting haemolymph around the mouth, hence the name. Oil-beetles mainly exude cantharidin in liquid form. This powerful poison is fatal to humans in doses as low as 0.03g; it also has a rapid caustic action on exposed human skin. The likely effect on the delicate mucus membranes of a predator's mouth can well be imagined. The great water beetle (*Dytiscus*) secretes a milky liquid from glands on the prothorax. The most active ingredient is cortexon, which induces narcosis in any fish unwise enough to include the beetle in its diet.

The European seven-spot ladybird, *Coccinella 7-punctata*, contains a potent combination of the alkaloids coccinelline and precoccinelline, which have a very bitter flavour. The blood, which contains the highest concentrations of these chemicals, is leaked to the surface through pores in the knee joints. It has a strongly deterrent smell of pyrazines, which are often associated with defensive odours in a variety of insects. Many soldier-beetles (Cantharidae), blister-beetles (Meloidae) and leaf beetles (Chrysomelidae) also practise reflex bleeding from leg-joints. Chicks of the blue tit, *Parus caeruleus*, are poisoned when fed seven-spot ladybirds, so it is in the best interests of both predator and prey to call a halt before ingestion actually takes place. A bad taste, strongly evident even after the merest touch with the mouth, will usually extinguish further

This seven-spot ladybird, **Coccinella 7-punctata**, is tethered to a leaf above the pupa of the braconid wasp, **Perilitus coccinellae** (see box on page 142 for a further explanation).

interest before any damage is done. Chemically protected insects tend to be a tough bunch anyway, well able to withstand a fair degree of punishment without being permanently harmed.

The risk that events will even reach the tasting stage is minimized by the strikingly memorable uniforms usually flaunted by such insects. The black-and-yellow stripes so familiar in *Vespula* spp. social wasps are a typical example of this 'warning coloration'. This constitutes a very simple 'keep off' message which serves both prey and predator well. Inexperienced birds, such as blue jays, suffer severe distress and vomiting not long after making a meal of a monarch butterfly, *Danaus plexippus*. Fortunately, most birds seem able to associate such purgatorial events with the meal which caused them, although how they make the correct connection between the two events is a mystery. The characteristic colour and pattern of the monarch thereafter invokes an aversion which lasts for a long time, probably for life. The same long-term aversion is also induced by warningly-coloured insects, which just taste bad, but would have no unpleasant effects if swallowed. Flavour alone is obviously as important to many animals as it is to us.

Many warningly coloured insects, especially in their immature stages, form dense aggregations, thereby drawing joint benefit in terms of predator avoidance. This situation does not always last into the adult state, which may be quite cryptic. The flamboyantly patterned nymphs of the coreid bug, *Pachylis gigas*, from Mexico crowd into dense groups. Yet the large solitary adult is basically dark brown. However, if molested, it turns and squirts the attacker powerfully in the face with an unpleasant anal discharge.

No warning colours, no matter how bright, can be perceived and avoided by a predator which hunts by night. Many tiger moths (Arctiidae) solve this problem by using a medium to which bats, their main predators, are supremely attuned – namely sound. When it detects the sonic pulses of a bat homing in on it, the moth signals its distasteful nature with a burst of ultrasonic squeaks from its thoracic vesicles. Upon receiving this warning, the bat usually breaks off the chase.

CHEMICAL MIMICRY

The diverse assemblage of insects which live as 'guests' in the nests of ants or termites ensure long-term survival in this hostile environment in a number of ways. Some stay alive merely by hunkering down when under attack by the nest's custodians and letting an armoured carapace take the flak. It would clearly be preferable to stay clear of trouble in the first place and a number of insects do in fact manage this through chemical mimicry. The peculiar slug-like larvae of *Microdon* spp., hoverflies, (Syrphidae) subsist on a diet of ant larvae and pupae. Despite their destructive inroads into this most vital of the nest's assets, the marauding larvae are free to move around and eat their fill without any reprisals from their heavily armed hosts. It seems that the *Microdon* larvae have cracked the ants' olfactory nest code by synthesizing the correct nest scent for the ant species which is their host. The ants therefore treat them as brood which has every right to be present. By contrast, the scarabaeid beetle, *Myrmecaphodius excavaticollis*, gradually acquires the nest scent over a number of days, eliminating any restriction to a single species of ants' nest. Staphylinid beetles in the genera *Zyras* and *Lomechusa* utilize yet another component of the chemical armoury. When molested by an ant within the nest, the beetles release a substance, which mimics the host's alarm pheromone, causing an outbreak of alarm behaviour. In the ensuing confusion the beetles have time to break and run for it.

Chemical mimicry is not confined to ant-guests (myrmecophilous insects). The tiny females of the parasitic wasp, *Lysiphlebus cardui*, are able to move freely around colonies of their aphid host, *Aphis fabae cirsiiacanthoidis*, despite the presence of cohorts of aphid-attending ant 'guards'. It seems that the parasite's cuticular secretions render it chemically invisible to the ants, whose tactile equipment is also plainly unable to distinguish parasite from aphid. In fact, the ants frequently palpate the wasp with their antennae, just as they do with an aphid when begging for the release of a honeydew droplet. Unchallenged in its cloak of

VISUAL MIMICRY

Most chemically protected insects are warningly coloured and often bear an easily memorized pattern. In a simple exercise in counterfeiting, numerous perfectly palatable insects (the mimics), gatecrash on this protection by copying their unpalatable fellows (the models). Such palatable fraudsters are called *Batesian* mimics. However, by all wearing the same 'club badge' the unpalatable models of many different kinds can also earn a defensive bonus, as a predator only has to memorize a single common pattern. In cases where actual ingestion is required to trigger further rejection, only a single club member has to be sacrificed in order to protect the rest. Such *Müllerian* mimics often mix with their Batesian copycats in complex mimicry rings. The situation is further complicated by the fact that the models themselves may vary in palatability, even within a single species, from extremely nasty to perfectly edible, depending on the toxicity of the food plant upon which the larva has fed. The whole system is based upon the important fact that birds usually need only a *single* unpleasant gastronomic experience to avoid further contact with a similarly coloured prey item, including of course, its mimics if these are good enough. Mimicry rings are based on a whole variety of models, including bees, wasp, ants and lycid beetles.

The slightest disturbance near a feeding ***Crinodes*** *sp. moth caterpillar (Notodontidae) causes it to flop its head backwards towards the source of trouble. This action reveals two eye-like spots between the last pair of legs. Combined with the rather fang-like effect of the foremost pairs of legs, this makes the caterpillar resemble a small snake about to strike. Costa Rica.*

*Nymphs of the praying mantis (**Phyllocrania illudens**) from Madagascar spend most of the day hanging head-downwards, resembling withered leaves. This silvery form corresponds with the bleached dead leaves typical of the dry deciduous western forests.*

The tropical rainforests of South America contain an extensive mimicry ring based on various common black and red wasps in the family Braconidae. Although the picture would appear to suggest that the subject is a fully-winged adult, this ***Aganacris*** *sp. katydid (Tettigoniidae) from Peru is in fact a nymph. The wing-like pattern along the back is pure deception, aimed at convincingly conferring the appearance of a fully-winged braconid wasp. The nymph also moves like its model, constantly shivering its black-and-white antennae. Even these come within the general specification for a perfect mimic, being amazingly short for a katydid, but much the same length as those of the model.*

Of the numerous wasp-mimicking hoverflies (Syrphidae) in the British Isles, by far the most plausible, is ***Chrysotoxum cautum.*** *Surprisingly enough, when birds were tested on a range of such mimics, the one which they appeared to shun most readily was* ***Episyrphus balteatus****, a species which few human observers would consider to be at all wasp-like.*

chemical anonymity, the parasite actually searches out colonies 'defended' by ants. It seems that these are less heavily visited by hyperparasites which, unlike *L. cardui*, lack the protection endowed by a suitable chemical veil.

CAMOUFLAGE

COLOUR

At its most basic, camouflage consists of a green insect such as a grasshopper spending most of its time on a green background of vegetation, or a brown bug living on the bark of a tree. In a slight increase in sophistication, insects which live on plants having numerous highly dissected leaves often have one or more pale stripes along the body, for example, the North American stick insect, *Timema tahoe*, which lives on the pine (*Abies concolor*). The stripes probably simulate the effect of bright light shining between the leaves, rendering the insect less conspicuous. When dissimilar colour forms occur within the same species, they may behave in a different way in order to maximize their chances of remaining undetected. Thus, the green form of the hawkmoth caterpillar, *Erinnyis ello*, rests among tree foliage, while the brown form prefers the trunk.

Insects which live on food plants having numerous small or narrowly dissected leaves often bear rows of pale stripes, resembling bars of light shining through the leaves. This tailed emperor butterfly (**Polyura pyrrhus**) caterpillar (Nymphalidae) is from Australia.

Most cryptic insects are very good at selecting the optimum background on which to spend the day. Unlike in many photographs, this mottled beauty moth (**Boarmia repandata**) was not collected at night and then artificially posed for photography, but was located resting on a pine trunk in an English forest. Note the vertical orientation, neatly fitting the body into a flaw in the bark.

SHAPE AND COLOUR

A major advance has been the evolution of forms which actually resemble some complete object in the environment. This must not, of course, be something which will be of interest to insectivorous enemies. Thus, there are insects which mimic leaves, both living and dead, twigs, flowers, fallen buds, seeds, stones and bird droppings. Some of the leaf-like tropical katydids (Tettigoniidae) are extraordinary for their authenticity. Not only is the ramifying arrangement of leaf veins faithfully reproduced upon the wing-cases but also the imperfections which would normally be present, such as the blotches caused by fungal attack, nibbled edges, white 'bird droppings' and even holes (not real holes, but transparent areas of chitin).

Living leaves tend to be regular in shape (bilaterally symmetrical). Dead leaves, by contrast, have a habit of becoming distorted as they dry and shrivel, losing their symmetry. Certain dead-leaf mimicking moths, such as *Oxytena modesta* from South America, overcome their own natural bilateral symmetry by curving their abdomen over to one side, rendering them closer to the real thing. The Madagascan praying mantis, *Phyllocrania illudens*, resembles a shrivelled leaf, with the leaf 'tip' being a twisted extension projecting upwards from the top of the head. For most of its life, this mantis hangs head-downwards

Mimicry of bird droppings is found in a range of both adult and larval insects. This **Euproctis conizona** moth is sitting in full view on a leaf in a Kenyan rainforest. Note how the spreading hairs on the legs resemble the way in which the liquid content of a dropping splashes outwards when it lands. When the leaf was disturbed by the photographer's movements, the moth immediately revealed that it had a back-up defence in reserve. It inserted the tip of its abdomen up through the wings, revealing a rusty-red warningly-coloured tuft of anal hairs containing defensive chemical secretions.

from a stem, where it looks like a leaf which has died but not yet fallen. In the cool rainforests of eastern Madagascar, where dead leaves remain brown in the humid environment, *P. illudens* is also brown. By contrast, in the hot dry deciduous forests in the west, where a fallen leaf is soon bleached silvery-grey by the sun, the mantis is coloured likewise. By always matching the preponderant leaf colour in their particular environment, the mantids decrease the prospect of being singled out as frauds.

Many stone-mimicking grasshoppers show similar local variations, on black volcanic ash, the grasshoppers are black, but are white on nearby limestone.

CAMOUFLAGED GRASSHOPPERS WITH A SPLASH OF COLOUR

When they are disturbed, many grasshoppers which blend in well against bare stony ground unfold unexpectedly ostentatious red, blue or yellow wings and take to the air. This sudden and unforeseen flash of colour no doubt startles a predator, who then makes the mistake of searching the grasshopper's landing area for a patch of bright colour, but by then the grasshopper has closed its wings and again merges perfectly into the background.

ACTIVE DISGUISE

Some insects implement the art of active disguise. Assassin-bug nymphs, which stealthily stalk their prey on the disintegrating surface of dead trees, adorn the whole body, legs and all, with a confetti of shreds and scraps of wood. The material stays in place by adhering to special glue-secreting hairs, making the bug look like nothing more gastronomically inviting than a mobile wood-chip. Some African dung beetles use their middle legs to ladle liberal quantities of soft fresh dung up onto their backs, where being fairly sticky anyway, it tends to stay in place. *Eurychora* spp. ground-dwelling tenebrionid beetles camouflage their backs with a close-packed covering of sand grains, while some moth caterpillars clothe themselves in a dense overcoat of their own droppings, ending up as a less-than-appetising meal.

Caterpillars of South American *Prepona* spp. butterflies spend the daylight hours suspended head-downwards from the mid-vein of a leaf, where they resemble a brown tip which has shrivelled and died. The caterpillar makes careful preparations for its deception by first chewing off the genuine leaf-tip. It then devotes many hours under cover of darkness to extending the mid-rib downwards with a silken strand woven from the labial glands. At the base of this extension-piece, where it adjoins the real leaf, the artisan carefully inserts some bits of real leaf which it has bitten off. These quickly turn brown, forming a convincing lead-in to the counterfeit section comprising the caterpillar itself, whose shape also resembles a ragged dead leaf.

DEFENSIVE DISPLAYS

No matter how convincing the camouflage, some cryptic insects are eventually going to be spotted for what they are by an inquisitive predator. When this crunch-moment arrives, all is not inevitably lost. It may be that the insect is chemically defended anyway, or can fight back in other more painful ways but trusts to crypsis as a primary tactic to stay out of trouble. Its warning colours are tucked out of sight, reserved for use in a tricky situation. For example, *Dirphia* spp. moths normally spend the day resting

This male **Neptunides polychromus** reacts to a threat by opening its spiny front legs in a defensive stance. This is obviously of little use when faced with a powerful and fearless enemy such as a civet, whose droppings are often speckled bright green, crammed with the undigested chitinous remains of these beetles.

inconspicuously among vegetation. If touched, they abruptly raise their wings and curve the abdomen downwards, revealing the densely hairy upper surface with its distinctive warning pattern. The Australian katydid, *Acripeza reticulata*, performs a similar display, while some of its equally distasteful relatives abruptly reveal brightly coloured legs or the undersides of the wings or wing-cases.

If a cryptic insect also happens to be edible – and the majority probably fall into this category – it may have to gamble on a display based on pure bluff to get itself out of a tight situation. The most frequent of these phoney performances is suddenly to reveal a pair of large eye-spots on the wings or the undersides of the wing-cases, or even, as in some mantids, on inflated areas on the front legs. This has the additional effect of abruptly increasing the apparent size of the displayer, confronting the enemy with what seems to be a large staring pair of eyes, possibly of an animal large enough to pose a threat. Monkeys and birds have both been seen jumping back in alarm when exposed to such performances, which are typical of many large moths (for example, *Automeris* spp., Saturniidae), some katydids (for example, *Tanusia brulatti*), mantids (for example, *Pseudocreobotra ocellata*) and the spectacular peanut bug, *Laternaria laternaria*. Under natural conditions, the insect waits until actually touched or grasped before launching into its act. In theory, if it is dropped in surprise, it has a good chance of making a complete escape. The caterpillars of some hawkmoths (Sphingidae) draw in their head when alarmed to bring into frontal prominence two or more eyespots. By thrashing their heads from side to side, sometimes accompanied by a distinct squeaking sound, they resemble a small and very intimidating snake.

A LADYBIRD AS A BODYGUARD

The larva of the braconid parasite, *Perilitus coccinellae*, guarantees the continued presence of its former ladybird host by imposing purely physical restrictions. After boring its way out through the underside of the ladybird, the larva fastens it to the leaf with silk and pupates beneath its body. Despite its injuries, the ladybird does not die but remains tethered in place, often struggling to break free, yet all the while doing involuntary duty as a warningly coloured shield protecting the vulnerable pupa of its enemy beneath.

PERSONAL BODYGUARDS

Personal security forces have been recruited by certain insects for a very long time but, as always, there is inevitably a price-tag attached for 'services rendered'. Probably more than half of all the known species of butterflies in the family Lycaenidae (the 'blues', etc.) have developed a relationship with ants. Some of these are purely antagonistic, with the lycaenid larva feeding on the ants' brood. Others are more give and take, with both partners deriving some benefit from a mutualistic association. In such an alliance the price payed by the larva to its ant guardians is a nutritious liquid produced from a special gland on the top of abdominal segment VII. The ants stroke this area with their antennae, upon which the sought-after liquid is served up. However, its manufacture is at the expense of nutritional reserves which could otherwise be used in growth. Larvae which have to produce a continuous supply of liquid to keep their defenders at action stations turn into rather smaller adults. In order to keep this cost within bounds, the larva gears its output to circumstances. If danger threatens, it instantly steps up production, quickly recruiting more ants to form a cordon around its body. Quieter times allow reduced output, just sufficient to keep the ants interested.

The larvae of the parasitic wasp, *Cotesia glomerata*, seem to modify their host's behaviour, so that it continues to act in their best interests once they have bored their way out *en masse* and pupated around it. During this vulnerable emergence stage, they induce temporary paralysis so that their host does not make any sudden movements which would dislodge them. They also ensure that it does not bleed to death through the numerous exit holes by leaving behind their cast laval skins as plugs. Instead of now wandering off in a vain attempt to pupate, as would be expected, the caterpillar stays put and lays down a silken sheet over the pupae; this helps to protect them from hyperparasites. When molested, the caterpillar thrashes around and regurgitates gobs of red liquid, which would probably be fatal if smeared on a tiny hyperparasite seeking to gain access to the crowded wasp pupae. By controlling the behaviour of their former host and causing it to remain protectively over them, the wasp larvae pre-empt its normal defensive repertoire for their own advantage.

Ant-guards commonly attend colonies of homopteran bugs, as well as the larvae of many lycaenid butterflies. The attraction in these **Aethalion reticulatum** nymphs in a Peruvian rainforest is the copious supply of honeydew on tap. The ant at top-centre is palpating the rear end of a nymph, stimulating the emission of a sugary droplet.

08 Insect Ecology

DISTRIBUTION

Insects may be found in almost any kind of environment from the most hospitable to the most hostile, as long as there is food available to them. They are found in polar regions and in the tropics, up mountains and in the depths of caves, from the arid deserts to the waters of ponds, lakes and rivers. The only environment that they have been unable to conquer is salt water, though a few species live along the shore below high water, surviving a twice-

The plant-sucking heteropteran bug, **Coreus marginatus**, is a member of the family Coreidae, which attains its greatest diversity in the warmer parts of the world. In the British Isles, it occurs only in the warmer southern half of England. Its nymphs are very cryptic, resembling the developing seeds of their **Rumex** host plants.

daily submersion by the sea. Giant water striders of the genus *Hylobates* even inhabit the Pacific Ocean, though being surface dwellers, they have not had to adapt to submersion in saltwater with its attendant physiological problems. A number of fly larvae are able to withstand high levels of salt in the water in which they live, notably those of the family Ephydridae, but they are exceptional. Fly larvae of this family live in salt-pans and salt marshes all over the world and *Ephydra hians* larvae survive in Mono Lake in California, which is much saltier than seawater.

The actual distribution of individual insect species may be dictated by a variety of different factors. The most important of these is the availability of food. Though there are insects which have very catholic tastes and are able to eat almost anything, most are restricted in their diet. Many herbivorous species rely upon just a single species of plant for their survival and most parasitic insects are restricted to a single host. Thus where the plant or host is found there the insect may also be found, though further factors, such as climate, may determine its presence. Take, for example, the European bug, *Coreus marginatus.* In the British Isles its nymphal food plants, members of the family Polygonaceae, including the common dock species, are ubiquitous. Despite this wide availability of its food plants, the bug itself has a very southern distribution and even then it is patchy, being at its most abundant in the very south of its range, in for example Cornwall, where in some years it can be very abundant. The family Coreidae as a whole tends to be insects of warmer climes and this fits in well with the fact that, as well as *C. marginatus,* all of the remaining British coreids also have a southern distribution.

Whereas the coreid bugs tend to be mainly tropical and subtropical in distribution, there are other groups of insects which show the reverse. Bumble bees, for example, tend to be much more common and show greater speciation in temperate regions of the world. Even within the bumble bees, however, there are extremes, with some species adapted for living on the slopes of the Himalayas while others inhabit tropical forests.

How then are insects adapted for living in one particular ecological niche? We will consider the desert environment as an example, though tropical forest or freshwater would do equally well to show the sort of adaptations involved.

DESERT INSECTS

A desert can roughly be defined as a region which receives 250mm or less of rain annually. How this rain occurs may vary from one desert to another. It may, for example, rain once a year over a short period producing a sudden rush of growth of annual plants, which must grow, flower and set seed before the ground dries out again. In other deserts, rain may be sporadic, with perhaps a number of years passing between each appreciable rainfall, which, when it comes, will again promote a luxurious growth of annuals. Some deserts are hot, with air temperatures as high as 50°C and ground temperatures as high as 80°C during the day. Other deserts are cold, with moderate day temperatures, dropping to below freezing at night. With such variations it is small wonder that insects find life in the desert difficult but, being tenacious beasts, many have overcome the problems and are able to thrive under these extreme conditions.

There is no one particular group which can be associated with deserts alone, though there are some, such as beetles of the family Tenebrionidae, which have many species adapted to a desert existence. On the other hand, there are insects to be found in deserts which we would hardly expect to find there, since they depend on water for at least part of their life cycle; dragonflies and damselflies come immediately to mind. What sort of adaptations are found in these insects which enable them to survive desert conditions?

What leaps to mind immediately is, how are insects going to obtain water when, for most of each year, none is available? One way is to only be around when water is around, thus many insects go through their whole life-cycle during the few weeks that food is abundant following the seasonal rains. At the end of this period, one stage of the life-cycle,

A group of insects which seem particularly well suited to life in the world's deserts are the beetles of the family Tenebrionidae. Their catholic tastes allow them to make use of almost any living or dead organic matter as a food source. **Sepidium tricuspidatum,** seen here in Israel's Negev Desert is one such beetle. The white pile covering its body and appendages is useful in two ways: it camouflages the beetle, hiding it from its enemies against the pale desert soil, and the white surface reflects the heat from the fierce desert sun.

whether it be egg, larva, pupa or adult, must then be adapted to survive in some way until the next rainy season. Water-dwelling insects have similarly to get through their breeding cycle before temporary ponds dry out. The dragonflies and damselflies already mentioned, which appear to be out of place in the desert, may in fact breed in the more permanent rivers and streams, which do flow through desert areas, perhaps fed from melting snows in surrounding mountains. For those insects which are active outside the wet season then other adaptations are necessary. Many are physiologically adapted so that they are able to obtain sufficient water from their food, coupled with losing very little water in the form of urine or through their body surface. The latter may be inseparable from behavioural adaptations, whereby such insects emerge to feed and search for mates only at night when it is cooler and they will lose less water.

Even in the non-rainy parts of the year, water may condense from the atmosphere during the night and many insects are able to avail themselves of this. Much of the water in the Namibian Desert of southern Africa arrives in the form of thick mist, blown inland from the cool water of the Atlantic Ocean. Some tenebrionid beetles adopt a posture which permits this mist to condense on their body surface and then run down to their mouth to be imbibed. Yet another way of overcoming the lack of water is to store it in some form when it is abundant. This strategy is adopted by the so-called 'honeypot' ants, which have a caste called 'repletes' adapted as storage vessels. The repletes hang from the roofs of tunnels in the nest and during the wet season they imbibe large quantities of sugar solutions, brought in by other worker ants, into their crops. Their abdomens become enormously distended with this 'honey', which then supplies the whole colony with both food and water when these are in short supply outside the wet season.

Many perennial plants, such as cacti and succulents, creosote bush, sagebrush and numerous others are themselves adapted to desert life, thus they provide a constant source of food for any insect that is capable of feeding on them. The word 'capable' is important, because many desert perennials contain toxic chemicals aimed at preventing their destruction by herbivorous animals. Consequently, any desert insects have to have their internal biochemistry and physiology adapted so that they can counteract the effects of these toxins. It is not usual to find such insects feeding during the hottest parts of the day, this process usually being restricted to early morning, late evening or the night. Desert katydids, for example, may be found deep down at the base of agave leaves during the hours of daylight, emerging only at night to feed. Desert crickets also feed at night, staying in their underground burrows during the day. One way to avoid desiccation is to live not on but in your food plant. This habit has evolved in the larvae of the North American giant skippers (Megathymidae). Their alternative name is yucca skippers and the caterpillars feed and grow within the succulent tissues of yucca and agave plants. Even the adult megathymids are desert adapted flying only during the twilight hours. In both North and South America, cactus plants may be found collapsed and rotting, often riddled with holes. This is the work of the larvae of jewel-beetles (Buprestidae), which burrow and feed within the cactus tissues. The organ pipe cactus of southern Arizona suffers in this way, the holes created by the beetles serving as a home for a species of desert earwig.

From the forgoing, it can be seen that desert insects adapt their behaviour to suit their environment. They tend to time their reproductive phases to coincide with the rains and times of plenty. Many have resting phases, which avoid the hottest and driest times by going into a state of suspended animation (diapause). Whatever stage of its life-cycle goes into diapause, the insect seems to be able to predict the beginning of the next wet period and emerges accordingly. In those deserts where rainfall may not occur at all for several consecutive years, insects are able to continue their diapause until such a time as it does rain.

Many, though by no means all, desert insects avoid the heat of the day, remaining within burrows or in the shade of plants at this time and emerging only when it is cooler, in the evening, early morning or at night. This also confers on them a further advantage of being able to avoid the attentions of day-active desert birds and lizards. If insects have to be active during the day then they may show other compensatory behaviours or adaptations. Insects which live in sand deserts, for example the American desert cockroach (*Arenivaga investigata*), effectively swim beneath the surface through the loose sand grains. Sand desert beetles often adopt similar strategies as do, interestingly, a number of desert lizards. Many diurnal desert-floor-dwelling beetles are either black or white in colour. Black may not seem a good colour to be, but remember that although black objects absorb heat well, they also radiate it just as well. White beetles, on the other hand, will tend to reflect heat. Many of these beetles are long-legged, some extremely so and this enables them to hold their bodies away from the hot desert floor as they go about their daily round.

DISPERSAL AND MIGRATION

One way in which an insect can overcome problems, such as climatic extremes or temporary lack of food, is to move. This may involve just a local dispersal or a much longer distance mass movement, a migration. The distances involved in such movements vary from just a few to hundreds of kilometres. Aphids, for example, in late summer and early autumn, disperse from their main summer food plants to the plants on which they overwinter, travelling relatively short distances, perhaps just a few hundred metres. American monarch butterflies, however, can travel hundreds of kilometres between their overwintering sites in Mexico and California and their breeding areas over much of the USA. Similarly, adult ladybirds (ladybugs) of certain species, in both Europe and North America, migrate into high mountain areas where, like the monarch butterflies, they hibernate in congregations of millions. Such large numbers of these insects in one place might seem to create a dangerous situation but both ladybirds and monarchs are distasteful and warningly coloured to put off potential predators. In the case of the monarchs, however, some jays have learned that a small percentage of the butterflies are actually edible, so they pick them up and taste them, discarding the inedible ones and eating the rest. This is unfortunate for the butterflies, since large numbers of those tasted are fatally injured.

Not only do adult insects migrate to suitable hibernating sites, they also migrate to escape the rigours of the hot, dry seasons of the tropics. In Australia, for example, caterpillars of the bogong moth, *Agrotis infusa* (Noctuidae) feed on the luxuriant winter grassland of northern New South Wales and southern Queensland, eventually pupating inside a cocoon in the spring. They eventually emerge in high summer, when the grasslands are dry and brown, totally unsuitable for beginning the next generation. Consequently, the adult moths migrate southwards to eventually aestivate (avoid the heat and dryness of summer) in the cooler Australian Alps. Here, countless thousands of them congregate in caves and on the surface of rocks, remaining immobile until the autumn. By now, the rains have started in the grasslands further north, so the adult moths begin their trek back to where they were born, to mate, lay their eggs and eventually die.

Many insect dispersals relate not to changes in season but to pressures induced by sheer numbers, as populations build up when food is plentiful. The most often quoted example, because of the havoc it wreaks, is the migratory locust, *Schistocerca gregaria.* It is resident over much of northern Africa, the Near East and into India. In a good year, when plenty of food is available, populations in some areas can increase to many millions. The result is that, as food supplies begin to run out, the insects go into a migratory phase, which goes in search of new food supplies. The insects are somehow able to detect areas of low pressure where rain and thus green food is likely to be present. Flying adults lead the swarms followed by the, as yet, unwinged nymphs, called hoppers, on the ground. They eat anything in their path that they consider edible, devastating large areas of crop plants as they do so. A number of other large grasshoppers, which is all locusts actually are, cause similar problems in some years in other parts of Africa, in Australia, South America and even in the USA.

INSECTS AND MAN

With so many different species of insect in the world and so many humans to come into contact with them, it is inevitable that they are going to effect one another's existence. We, however, can only really look at the man–insect relationship from our point of view, that is whether it is to our advantage or to our disadvantage, or perhaps just neutral. Advantageous insects we tend to nurture, disadvantageous ones we destroy, neutral ones we usually just ignore. It is, however, those insects which we consider to be our friends and those which are enemies which concern us here.

USEFUL INSECTS

The honey-bee is the best known useful insect. Whether they are wild in the world's forest or maintained under the controlled conditions of a hive, this insect provides us with a delicious food in the form of honey and a useful product in the form of beeswax, while many swear by the rejuvenating effects of royal jelly.

Not only does the honey-bee provide for us in this way but it is an important pollinator of many of our cultivated plants. There is, however, a down side to

The honey bee, **Apis mellifera**, despite the fact that it stings us sometimes, is always thought of as a great friend to man. This particular individual is pollinating apple blossom in the author's garden, the result being a very good crop of apples (*right*).

The katydid **Ruspolia differens**, here lying well camouflaged on the leaves of a doum palm in Kenya, is greatly prized as a food item in East Africa. They are nocturnal in their habits and on wet nights thousands of them are attracted to lights in the towns, enabling large numbers to be collected by the local populace. Six different colour forms of this katydid are known, matching them to the particular kind of grass upon which they are feeding.

INSECTS AS FOOD

Most of us in the so-called 'developed world' consider honey, formed from regurgitated nectar, to be very tasty and probably eat food coloured by cochineal without realizing that it is extracted from the body of a sap-sucking bug. Few of us, however, would be prepared to eat the actual insects if offered them, despite the fact that we eat other arthropods (crabs, lobsters, shrimps and prawns), and consider them to be a luxury. Yet insects are a rich and renewable source of protein and they form an important part of the nutrition of a number of peoples from around the world.

Insects, in fact, provide ample protein, as much as lean beef, although its amino acid content is not as well balanced as beef for human nutrition. Luckily, those amino acids which are in short supply can be made up from easily available plant proteins such as those found in cereal seeds. Insects are also a relatively good source of minerals and some vitamins.

So, who eats insects and what sort of insects do they eat? A well-known example is the 'witchetty grub'. This large cossid moth larva, which lives inside and feeds on acacia stems and roots, is dug out by Australian aborigines and eaten immediately with great relish. Historically, the same people collected vast numbers of bogong moths (the noctuid *Agrotis infusa*) from their aestivating sites and then baked and ate them. In Africa, in years when they are particularly abundant, hundreds of tons of the 'mopani worm', the

A plate full of fried katydids, **Ruspolia differens** from East Africa, ready for human consumption. The wings and legs are removed before they are cooked.

caterpillar of the saturniid moth (*Gonimbrasia belina*), are eaten by the indigenous population. Unlike the witchetty grub, which is eaten whole and raw, the gut has to be removed from the mopani worm and it is then baked before being eaten. Still in Africa, especially Angola, but also in other developing countries, the larva of the palm weevil, *Rhynchophorus phoenicus*, is an important source of protein and fat. However, on the other hand, the palm weevil is an important pest in cultivated palm plantations. Other examples of insects that are regularly eaten include grasshoppers and silk moth pupae in Korea, cicadas in New Guinea and giant water-bugs in South-East Asia. Insect eating is also widespread in South and Central America, especially in Mexico. In fact of the insects eaten by man worldwide 40 percent of the total are eaten by Mexicans. The list includes Hemiptera, Hymenoptera, Orthoptera, Coleoptera and Lepidoptera.

Whether insects will ever become more widely acceptable as human food depends upon many factors, some within and others outside our control. There is no doubt, however, that it would be ecologically sound to rely more upon insects as a source of protein food. It has been estimated, for example, that whereas one hectare of typical ranch land in the USA supports 100 kilograms of cattle, it supports 1000 kilograms of arthropods. The conclusions from these figures are obvious, when it is considered that we still cut down millions of hectares of forest to produce land for grazing cattle. Even if insects have been accepted as food by the majority of the population there are, however, many problems to be overcome if we are to farm insects instead of cattle. Insects are the prey of many other creatures, their numbers tend to fluctuate from year to year due to disease and parasites and they are small and therefore difficult to harvest and handle.

One way around these problems is to feed the insects to chickens. In India, for example, the by-products of silk production, in the form of de-oiled silk moth pupae, are fed to hens. Thus even if we do not relish the idea of eating insects directly, at least we can do so indirectly.

introducing bees into new areas as pollinators of crop plants. They are so successful that they often drive out native species of bee, which are themselves well adapted for pollination of native plants. The honey-bees are not as efficient in pollinating the latter and reduced seed set can eventually result in a decline of many native plant species along with the other plants and animals associated with them.

The honey-bee is an example of an insect which we deliberately nurture due to the benefits that we gain from it. However, there are many other insects which act as pollinators and which we consider as being beneficial – most species of bees, many wasps, butterflies and moths, beetles and many species of fly are all examples.

Insects may be productive in other ways. Despite the introduction of many synthetic fibres, silk, produced by the caterpillars of the silk moth, *Bombyx mori*, is still very much in demand for the production of high-quality garments. A plant-sucking bug, the scale insect (*Laccifer lacca*) from Asia provides us with shellac. Another scale insect, the cochineal bug (*Dactylopius coccus*) from Central America, yields the food dye cochineal.

INSECT PESTS

The idea of insects being pests is a human concept, for these creatures, like man himself, are only trying to ensure their survival in a highly competitive world. Unfortunately for the insects concerned, we are often interested only in our own survival and if they are pests, then they are to be treated as such.

DISEASE

Insect pests fit into a number of categories. There are, for example, those which affect humans directly either by being an uncomfortable nuisance or by carrying disease, though many do both. Fleas irritate by biting and they can also carry diseases such as bubonic plague. The same can be said of mosquitoes, which carry malaria and yellow fever and flies, which carry diseases such as sleeping sickness. Some insects can make our lives an absolute misery to the extent that they can virtually exclude us from

A number of insects undergo changes as population pressures increase, leading them to a dispersal phase when they go in search of fresh food sources. An example of this is the African elegant grasshopper, **Zonocerus elegans**, which can reach pest proportions in some years since it feeds on a number of crop plants. When plenty of food is available and dispersal is unnecessary, the adults have only tiny wings, no use for flying. This has the advantage that the energy that would have been put into making a set of full-size wings can be put to better use, making eggs for example. As populations enlarge and food begins to run out fully winged individuals (illustrated here), which can fly to new food sources, are produced instead.

a particular region. The far north is one such area, where, from mid-summer onwards, biting midges (Ceratopogonidae) occur in countless millions along with numerous black flies (Simuliidae).

FARMING

A second category of insect pests includes those which affect domestic animals and crops. Most of the insect pests which affect animals are carriers of disease, though biting flies can be an extreme nuisance to horses and cattle when they occur in very large numbers. Insects such as aphids and other plant-sucking bugs are likewise carriers of plant diseases.

Crops are, of course, affected in a different way in that many of them are actually eaten by insects. The devastating results of locust swarms have been documented on many occasions but other insects can be equally injurious. Moth caterpillars, for example, can defoliate vast tracks of commercial timber stands in a very short time with a subsequent reduction in the trees' growth. It is interesting to reflect, however, that all that these insects are doing is taking advantage of vast quantities of suitable food, which we have provided for them by having such large areas containing just a single plant species (monocultures).

STORED FOODSTUFFS

Insects can also feed on stored foods, and such material as leather, wool and the wood of buildings and furniture. Insects which take advantage of the vast concentrations of stored food include flour beetles, grain weevils and the larvae of various species of small moths. Wool is attacked by the caterpillar of the well-known clothes moth and timber is attacked by insects such as the death watch and woodworm beetles, as well as by termites in warmer climates.

A number of insects are considered to be pests if they are in the right place at the right time, that is if their presence coincides with that of man. Typical examples are social wasps and ants. Out of contact with man they play an important biological role in disposing of all sorts of waste plant and animal material as well as, especially in the case of wasps, destroying the larvae of a number of other insect pests. On the down side, wasps have an unpleasant venomous sting, which can be dangerous to sensitized individuals. Ants can also be a problem to man and soon attract the attentions of a dose of pest control officers. A further problem with ants is their liking for aphid honeydew. The result of this is that they often protect, and thus stimulate, an increase in the population of pest aphids.

BIOLOGICAL CONTROL

Biological control involves using an insect, not a pest, to control the numbers of another insect which is a pest or to control the numbers of a plant pest. The use of biological control to reduce pest numbers, rather than the accepted norm of insecticides, can have advantages. Most, if not all, insecticides and herbicides are expensive and they are to a greater or lesser extent toxic or produce toxic by-products in the environment. They also do not just target the pest insect or plant, harmless species often being wiped out along with the pest. Insect controllers do not have these side effects, as long as they are carefully chosen so that they themselves do not become pests. Their one disadvantage, which is perhaps only

A common pest of plants in houses, conservatories and greenhouses is the glasshouse mealybug **Pseudococcus affinis**. Heavy infestations, such as this one on a passion flower vine, can be highly damaging to the plants on which they live and under commercial conditions can render them unsaleable, thus they are of considerable economic importance.

KILLER BEES

All to often, when man interferes with the natural order of things in order to satisfy his insatiable appetite for more and better products, things can go wrong. One such example, which has tended to hit the headlines over the past few years, is that of 'killer' or 'Africanized' honey bees in the Americas. The story of these notorious insects begins back in the 1950s in Brazil. Apiculture in Brazil at this time was not very successful, partly because the strains of European honey bees used in most other places were not suited to the country's tropical climate. Wild colonies seldom survived and those in hives only thrived with considerable efforts from the beekeepers.

We now know that the failure of European bees to thrive in Brazil relates to factors controlling reproduction and increase in colony size. The domesticated honey bee is of Asian origin and evolved in conditions where increasing day lengths heralded the onset of spring and then summer, with an accompanying burgeoning in the number of flowers available as food sources. Unfortunately, in Brazil, changes in day length do not necessarily coincide with an increase in flowering; in fact, the opposite can be true. Thus the stimulus for increase in colony size in Brazil can be at a time when there are virtually no flowers and the colony fails as a result.

It was decided, therefore, to introduce into Brazil honey bees more accustomed to living in tropical conditions. As a result, a number of queens of the African subspecies were brought in for initial study. Before these controlled studies had progressed very far, however, some queens escaped into the wild. Not long after this event, despite knowing little about the survival and habits of these bees in Brazil, queens reared from the initial introduction from Africa were distributed to Brazilian beekeepers. Conditions were now available for cross-mating between the already present European and the newly introduced African honey bee strains to produce what we now call the Africanized bee. It was soon obvious that the Africanized bees were much more at home in Brazil than the European varieties. When there were lots of flowers around the Africanized bees had the ability to split off new colonies, a young queen and a swarm of workers leaving to found a new one while the old queen remained in the original colony. Furthermore, they also had the ability to remove all food supplies from the hive and then leave it to seek better conditions in which to found a new colony elsewhere.

Once the Africanized bees were established in the wild in Brazil, they spread rapidly beyond the areas where beekeeping existed into those where there was none. It was in the 1960s, however, that the first warning signs concerning their behaviour became apparent. Africanized bees showed a much more intense defence of their nests or hives than did the original domesticated strains and this made them much more difficult to handle. Indeed, when disturbed, virtually every worker bee present in a colony would leave the hive and attack the intruder. Initially this problem meant that there was little increase in beekeeping as a result of introducing African genes! However, by teaching beekeepers how to handle these touchy bees and by training them to kill queens produced by the most hostile colonies, a thriving industry was eventually achieved.

If the story had ended here, then the original aims of introducing African genes into American bees would have been achieved with the minimum of disturbance to the status quo. The Africanized bees, however, continued to spread in all directions into other South American countries, maintaining both their aggressive characteristics and their tendency to swarm regularly as they did so. By 1977 they were well into Venezuela and Colombia, by the mid-1980s they had reached Mexico and, despite attempts to slow their progress, they entered the southern states of the USA in the early 1990s. It is expected that their migration northwards will eventually be limited by climatic conditions, though by this time the damage will have been done, with much of the United States' apicultural industry in a turmoil as a result.

The reason for all of this interest in the Africanized bees is of course because, as their alternative name implies, they are killers. It is estimated that in South and Central America they have been responsible for the deaths of at least 1000 people, as well as thousands of domestic animals, who have disturbed their colonies. By 1995 there was one confirmed death in the USA, of an old man who was unable to run away from a swarm when it attacked him. There will no doubt be many more for it is unlikely, at least for the present, that we will be able to eradicate these 'killer' genes from bee populations, even if we want to.

The photographer approached this swarm of Africanized honey bees in Brazil with some trepidation. One false move and they could have attacked, with potentially lethal consequences. Fortunately it was raining and they were not inclined to do so, enabling a series of shots of the swarm to be taken.

To the assistance of the grower or collector of indoor plants comes the predatory beetle, **Cryptolaemus montrouzieri**, which feeds upon nymphs and adults of the glasshouse mealybug, **Pseudococcus affinis**. Here we see biological control, with the beetles being introduced onto a passion-flower vine heavily infested with the mealybug. The beetles are specially bred for this purpose and cultures are relatively inexpensive. They breed fairly rapidly and within a few weeks the adults and larval beetles are able to bring the mealybug infestation under control.

a small one (since even using pesticides, it is perhaps morally wrong to make any pest extinct), is that they do not eradicate all of the pests, since they will want further generations for their offspring to feed on.

Historically, one of the earliest examples of biological control using an insect was that of the prickly pear cactus in Australia and South Africa. Following its introduction, it adapted to its new countries so much that it quickly spread, forming dense, impenetrable masses where formerly there had been grazing land. It was finally controlled by the introduction of a pyralid moth (*Cactoblastis cactorum*) from South America, whose larvae specialize in feeding on opuntias. The larvae bite into the opuntia pads to feed, not only harming them directly but also indirectly, by allowing the entry of infectious agents into the internal tissues of the cactus.

More often, however, insects are used to control other pest insects. Two species, which can be bought off the shelf in the British Isles for example, are used to control the glasshouse mealybug, *Pseudococcus affinis*, a pest of many species of house plant. The first of these is a predatory beetle, *Cryptolemus montroverleri*, whose adults and larvae feed both adult and nymphal mealybugs. The second is a tiny parasitic wasp belonging to the genus *Leptomastix*, which lays its eggs inside the mealybug. The wasp larvae then feed on the mealybug's internal organs, eventually killing it. Both insects are quite successful at controlling the pests though, of course, they do not eradicate them. Regular reintroductions of the controlling insects into the greenhouse are usually necessary to maintain adequate control of pest numbers.

Man and insect have an uneasy relationship: we admire a few that are beautiful, praise those which help us, curse those that molest us and ignore the rest. The insects, however, inhabited the Earth long before we came upon the scene and they are likely to still be around long after we have gone. For even if in our folly we eventually make the world uninhabitable for ourselves, the insects, with their enormous adaptability will still soldier on in one form or another.

GLOSSARY OF BIOLOGICAL TERMS

Cerci Feeler-like extentions on the end of the abdomen.

Crypsis Camouflage coloration.

Ecdysis Moulting of the old exoskeleton.

Elytra Hardened front wings of a beetle that protect the delicate flying wings.

Haemolymph The usual biological term for insect blood.

Halteres Reduced stumps of the hindwings of flies important for balance.

Imago Alternative name for an adult insect.

Instar Pre-adult stage in the life-cycle of an insect.

Integument Tough outer layer of the insect making up the exoskeleton.

Lek Area in which a number of males gather, each adopting a small territory within it, the purpose of which is to attract females.

Metamorphosis Development changes during the insect life-cycle.

Nymph Pre-adult stage in the life-cycle of an insect.

Ocelli Simple eyes used to detect light and dark only.

Ovipositor Egg-laying tube of a female insect.

Parthenogenesis Development of an individual from an unfertilized egg.

Pheromone Chemical that attracts one sex to another, which may be airborne or smeared on a substrate.

Proboscis Sucking mouthparts of an insect.

Pronotum Hard shield overlying the thorax.

Rostrum Piercing and sucking mouthparts, particularly applied to bugs.

Sclerotized Hardened or toughened, referring to the exoskeleton.

Spermatophore Sperm package produced by the male and picked up by the female.

Spermatophylax A supply of nutrients attached to the spermatophore.

Stemmata Simple eyes found only on the head of insect larvae.

Stridulation Production of sound by male insects to attract females.

Viviparity Production of fully-developed young rather than laying eggs.

INDEX

References in *italic* refer to illustrations.

Index